THINKING OUTSIDE THE BOX

Thinking Outside the Box

The Most Realistic Way of Thinking, Adopting and Leading Life

We are born free, so hold head high, seek knowledge, and ask questions of any sort until answers are meaningful and justified. Fear, inaction, and silence are no service to humanity.

Abdur Rahim

Library of Congress Control Number:		2017900069
ISBN:	Hardcover	978-1-5245-7389-8
	Softcover	978-1-5245-7388-1
	eBook	978-1-5245-7387-4

Print information available on the last page.

Rev. date: 02/13/2017

To order additional copies of this book, contact:
Xlibris
1-888-795-4274
www.Xlibris.com
Orders@Xlibris.com
750817

CONTENTS

To the people, since ancient to modern times, who dedicated
their entire lives promoting and establishing a very simple way of life
—*freethinking*

Looking Beyond . . .

My inquisitive mind doesn't vibrate inside the fenced space;
An unfathomed box filled with false words of promise,
Awards and punishment, opium of the cowards;
No scope to think but to believe designated books.

So I walk around the open sky and see the world, the real world,
Searching for knowledge of all sorts beyond books and boundary.
That's why I seek teachings from the sky, what's in there.
I get lessons from the green, the taste of life.
I receive love, adoration, sorrow from women's laugh and cry.
Birds teach me the sense of liberty and freedom.
From folk singers, musicians, farmers, peasants, and workers,
I learn devotion, respect, labor, and passion.
Alas! In the closed confinement, I find no god or knowledge.
Alas! I am not capable of adoring mysticism of cult
In my thoughts, imaginations, and in every disposition of life.

But I find God's identity in all cells, particles, and matters,
Infinitesimal or vast, animate or inanimate,
And I see knowledge exists all over the nature.
That's why I search it in a much bigger horizon and stay outside
The surrounding walls of mosque, church, synagogue,
Or monks or any particular place of worship whatsoever,
Because these are places nothing but of hollow sounds.

—Abdur Rahim

FOREWORD

I have known the author Abdur Rahim for decades and am honored to have been invited to present some forewords for this book. Given Mr. Rahim's background as a specialist in statistics, the lucidity with which he handles topics of religion, humanism and transhumanism has certainly impressed me. In my professional life as a practitioner in development and an academic in development studies, I consciously avoid pursuit of discussions on religion, adhering to the principle of separation of state and religion. Humanism, of course is at the center of development studies and practice. Mr. Rahim's observations on organized religion and humanism have certainly educated me.

Discussion on religion and humanism, presented in the book, stimulates interesting debates, particularly relevant in the current tumultuous and insecure global context of conflict, fear, frustration and uncertainty. Mr. Rahim introduces readers to the concept of religion originating in human search for security against extreme hardships and dangers arising from natural disasters, unpredictable harvests and other uncertainties. He boldly asserts that blind faith in religion, holy scriptures and God offer a protective wall of comfort to followers of faith, with organized religion freely imposing rules, edicts and dogmas that believers must abide by. His analysis provides insights into the philosophy of humanism as a natural alternative to religion and reliance on blind faith, the dogmas and restrictions of organized religion. He argues that embracing humanism helps promote *humanity* (comprising compassion, understanding and mercy) as the primary driver for transition to a secure world and a better global society promoting better life for people. Humanism certainly plays a central role in international development, the field of my specialization.

The section on humanism leads readers to trace the historical development of humanism as a philosophy from the second millennium B.C. to the 20th century. The author's elucidation of the central principle of humanist philosophy- the use of reason and ethics for the betterment of humankind- will convince readers of the value of humanism. Similarly, readers are likely to agree with the author's view that humanism that cherishes tolerance and freedom of thought is not necessarily averse to all religions. However, it is difficult to gauge readers' reaction to Mr. Rahim's rejection of organized religion as a source of morality or law. He believes that morality is a natural and biological impulse, having little to do with religious edicts. Many readers while agreeing that ethics and principles that guide life can be set largely by individuals, might have difficulty embracing the notion of total rejection of demand of any organized religion for collective obedience to edicts. The author rejects the demands of blind faith, total obedience, and intransigence that he believes characterize religion. Thus, he issues an alert to "thinking outside the box," a proposal that will stimulate debates.

Mr. Rahim's claim that humanists believe that the universe and everything in it can be explained by natural laws, places great importance on science as the guiding factor in human life. Following Julian Huxley, Mr. Rahim presents transhumanism as a natural extension of humanism for enhancing human mental and physical capabilities by technological and scientific advancements. This section of his work will certainly raise interest and promote a dialogue. While transhumanists are optimistic of potential impact of advances in science and technology, it is hard for contemporary philosophers to predict the nature and types of technologies that may be generated decades or centuries in the future.

This is a thought provoking must-read book that promotes dialogue on primacy of reason, science and humanity over orthodoxy of religion. The book will be of help with your own journey toward a humanist destination.

Nipa Banerjee Ph.D.
Senior Fellow-School of International Development
University of Ottawa, Canada

Contract Professor
International Development Management
Sprott School of Business-Carleton University, Ottawa, Canada

ACKNOWLEDGEMENT

This book would not have been possible without the support and encouragement of many individuals. At the very outset, I must thank to Dr. Gul Hossain, Ex-Director of Bangladesh Agriculture Research Council (BARC), and prior to that he taught Genetics and Plant Breeding as a professor at the University of Calabar and the University of Nigeria, Nsukka. He not only provided valuable insights on the main theme of the book but also thoroughly edited the first draft of my manuscript. Furthermore, he put considerable time and effort to write a draft summary of the whole content of the book which helped me a lot to write the final version of it. His whole-hearted support and highly praising of my work ignited my inner energy to continue working on the manuscript. I am deeply indebted to him.

I am extremely grateful to Dr. Nipa Banarjee, Senior Fellow-School of International Development, University of Ottawa, Canada and Contract Professor, International Development Management, Sprott School of Business, Carleton University, Ottawa, Canada, for graciously writing the foreword for the book.

I would like to extend my special thanks to Dr. Maks Rahman, currently retired from Health Canada, who had always been with me along the long road to finish writing of the manuscript from the very early stage. At a time when I pushed my limit in editing the whole manuscript, Dr. Rahman eased my work by editing an important section, for which I would like to express my gratitude. Nonetheless, his consistent encouragement kept my mental strength up and going to finish the job I had taken on.

I am thankful to my long-time colleague Dr. Mark Hammer at the Public Service Commission (PSC), Government of Canada for his support and encouragement, who provided his valuable suggestions on the content structure and the flow of the theme. I am truly honoured by his sincere advice.

My sincere thanks extend to evaluators, copyeditors, and the staff at Xlibris Publishers whose professional and dedicated services, undoubtedly, helped to attain the book to a high standard in terms of quality, value, and marketing. Didi Rodrigues, Iris Johannsen, and Sam Clarke of Xlibris deserve special thanks for their good advice.

Last but not the least; I am not a scholar or a specialist, and I do not claim to be an original thinker rather than the fact that I have a passion for freethinking. So, for writing this book I have relied heavily on the valuable works of many scholars of various subjects, and have presented, as much as possible, the ideas and issues to the readers, what I have gathered from their works. I have paraphrased texts with proper acknowledgments, as needed, in the endnotes and bibliography. I have also quoted others' statements with proper authorship. Let me take from Ibn Warraq who quoted Goethe's advice to an author; "what is there is mine, and whether I got it from a book or from life is of no consequence. The only point is, whether I have made a right use of it." So, it is upto the readers to judge whether I have made right use or not. I am truely indebted to those authors whose works I extensively used in the book, as indicated at the endnotes and bibliography.

PREFACE

Thinking Outside the Box is the title of this book. By box here, I mean religion, not only in which people are affiliated but also which endorses a "must have blind faith" idea, beyond which they are not permitted to think freely and to search for real truth. This is synonymous to confinement into a closed box. But this should not be the lives of human beings. As humans, we are the superior species of all species on earth, and thus, we are free to think, gather new knowledge, innovate, and create new things by virtue of our brain's capability and power. This natural gift has made us always ask questions and attempt to provide answers to their unending questions. But religions block our freethinking and suggest not thinking beyond holy books but believing them blindly. Is this right? Is this fair? Is this good for human development? Is this human value and dignity? Is this humanity? I think not. But I hold someone said, "Belief is a potent drug that destroys the thinking abilities of the believers." The fact is, once believers become convinced of the truth of their religion, they justify everything, including lies. Generally, people with strong faith (who are usually decent and ethical) willingly lie to support their faith without any evidence or knowledge of the truth. The end (truth or not, known or unknown) justifies the means. French philosopher and mathematician Pascal wrote, "Men never do evil so completely and cheerfully as when they do it from religious conviction." History is the witness of the fact that Pascal said.

People get their brain from birth, and it is their birthright to think freely and to question anything that comes from the brain, including the text of holy books. But religions teach us not to think but to believe. You know one thing; if you do not teach your children to think, religions will teach them not to think but to believe. Remember, *it is easy to believe than to think*. I experienced this in my entire life, meeting with people of different religious affiliations closely and intimately. Nonetheless, religious faith is very strong in human minds; it does not die out from the brain, and it will not until human beings can overcome the fear of being insecure in the harsh and ruthless nature, the fear of death, and the fear of unknown and uncertainty.

There are two attributes that are importantly being adopted toward the unknown. One is to accept the pronouncement and hearsay of people who demand that they know, based on books, scriptures, mysteries, or other sources of information and inspiration. The other one is the practical experience gathered by going out and looking for by oneself. There are many questions that people think and ask themselves that religion cannot answer. We may ask questions like what is the meaning and purpose of life, if indeed any at all? Is the world designed by someone called God, if so at all? What is the purpose, if so at all? Are we, as human beings, made up of dust crawling helplessly on this small planet, as astronomers see it? Or are we made up of various chemicals mixing together in some intelligent way, as chemists say? Why these current set of laws, not the others? What could that be? It goes on and on. These are all puzzling questions indeed. To study this difficult subject and gather knowledge, we have to learn what others at different times have thought about these matters. As we come to understand from the wisdom of others, we can live better. The importance for us, as human beings, is to search the truth, and in doing so, one has to pursue knowledge. This has the context of ethical principle (which is strongly associated with humanity) that stems from Socrates and many other contemporary philosophers. But how can one take the ethical principle that the pursuit of truth is a good thing? Bertrand Russell argued this way: neither are we endowed with the ability to engage in the scientific enquiry nor is it possible to suspend judgment; we must act as well as think. Tolerance is the key, as a precondition, in a society in which inquiry to pursue truth is to flourish. Freedom of speech and thoughts and opinions is the significant promoter of a free civilized society, where everyone can pursue knowledge and search the truth. Everyone will not have the same opinion on everything and in belief systems in particular within a society, but it must ensure that no avenue is closed for skepticism. For us, the unexamined life is, indeed, not worth living. This is an intrinsic part of humanity.

The key theme of this book is to pursue humanism rather than to be bogged down blindly to any particular religion. Because humanism is a nonsuperstitious worldview that allows us to make more ethical choices based on rational desire to do the most possible good. All the world's major religions should unite the world population rather than divide them and should come together and promote one good human value called humanity. According to the Dalai Lama, the reality today is that grounding ethics in religions is not adequate to promote this theme. That is why he is convinced that the time has come to find a new way of thinking about spirituality and ethics beyond religion altogether. I think he meant humanity without religion.

God did not create man in his image. In fact, it was the opposite, which is the simple reason why so many gods and religions and so many killings of brothers and sisters both between and among faiths. Religious atrocities have occurred not because humans are evil but because of the fact that religions have made them irrational. I think it is not wrong to say that famous evolutionists or physicists or biologists are more enlightening, even when they are wrong in their works, than any person of faith who is vainly trying to explain how he (being a mere creature of the creator) can possibly know the creator's intents (Hitchens 2007). He recalled the works of astronomy and biology and said that in examining the symmetry of the double helix (when your own genome sequence is fully analyzed), you will be impressed to know the core of your being and reassured that you have so much in common with other tribes of the human species. Moreover, it would be more fascinating to learn how much you are a part of the animal kingdom as well. So you can be humble now to your maker, which is not to be a "who" but a process of mutation. Addressing all religious friends, Christopher Hitchens said, "Those who offer false consolation are false friends."

Religious faith is very strong and incredible among believers to begin with but not the end. It is not the end because humans are dynamic in nature. People change with the change of time; even their belief changes. Most of us were born in any of the religious family traditions. So in our childhood, we belong to a religion for sure. As we grow older and understand the world, we change, depending upon our life experience and the knowledge gathered from around us. At some point, some may think that religion is irrelevant to their lives. So people change and become freethinkers, atheists, agnostics, humanists, or whatsoever—any alternatives of religion. They are the outside of religious box thinkers. They think religions block freethinking; in fact, they do. Because religions teach us not to criticize, not to question, and not to believe any of the holy books other than the particular one that is yours. For example, Muslims will believe only Quran, Christians will believe only Bible, Jewish will believe only Torah, Hindus will believe only Bhagavad Gita, etc.

In the book, I have presented the case of transhumanism as the next advanced phase of humanism, which can happen only with the advancement of science and technology. But how would humanism be transcended to transhumanism? David Roden, in his book *Posthuman Life: Philosophy at the Edge of the Human* had defined transhumanism as a socio-ethical idea that the advanced forms of technology can be used to transcend certain limitations of the human condition. He categorically referred to NBIC technologies as nanotech, biotech, information technology, and cognitive science. While commenting on Roden's definition, David Pearce, a British philosopher, pointed out that transhumanists are committed to three things: super longevity, super intelligence, and super well-being. In other words, transhumanists are committed to using NBIC technologies to live radically longer lives, increase their cognitive abilities, and achieve higher states of conscious bliss and satisfaction. This view works very much within the humanists' ideology, because transhumanists are committed to enhancing and improving the kinds of attributes that humanists think are unique and special markers of humanity (that is, rationality, intelligence, autonomy, etc). They just want to do so through technology (Roden 2015). The above views have the relevance to the natural law of dynamism. In nature, everything is changing and evolving not only biologically but also technologically. In other words, both humans and the rest of nature are dynamic, and both are changing continuously. This is the idea that transhumanism transcends various ideas of humanism, because humans themselves evolve and change at certain rates.

Why have I included transhumanism? Because transhumanism is a human ideology driven by human interests, desires, and morals. It is a cultural and intellectual movement that is thought to improve the human condition through the use of advanced technologies. In pursuance of this movement, transhumanists are interested in inventing advanced technologies that can boost our physical, intellectual, and psychological capabilities beyond what humans are naturally capable of. That is why the name is transhuman. Although transhumanism is just a new term for a very old phenomenon that was practiced in the long past, people had been augmenting humanity with tools long before modern technologies came into effect.

I am one of the freethinkers. I was born in a Muslim family, and I was taught that Islam is a religion of equality, a religion of peace, a religion that encourages seeking knowledge and searching for truth, a religion that is based on the reality of life, and a religion that rejects violence. Teachings went on and on, but at certain points, my practical life experiences were added to my realization of what the religions (not only Islam but also all religions) were all about. To have comprehensive knowledge of religions, I started studying them, though not extensively but necessarily enough to have a handful of overall knowledge so that I can console myself and have control and lead of my life without religion. This does not mean that I am hateful of religions; rather, I did and still continue to respect people of all faiths and religions. But I am critical about the fact that, in return, my religious friends will leave me alone in their own way. You know what I mean? Yet being a freethinker, I do advocate that humanism is the better path for human beings to lead a good life without religion, and I am a strong proponent of humanistic ideology.

Let me raise a question, what is the guilt to anyone to become a nonbeliever of any religion and identify him as a humanist at a certain time of his/her life? What is your answer? I presume the answer will be a dichotomy; some will say, "No problem," and others will say, "Yes, there are problems." The "yes group" creates the problem, as they always see the nonbelievers in a very negative way, even if he/she is being humiliated socially or at the personal level of a relationship. He or she is not warmly welcome in the family or society, a kind of discrimination. But vice versa does not appear to be seen. A humanist does not look at believers in a negative way, but believers do to the humanists. This negative attitude appears to be pronounced among the believers of the major monolithic religions. Why is that? To me, it seems racist. On the contrary, humanists are generous people who possess high ethical values and morals than their believer counterpart. The obvious reason is that the humanists promote love, friendship, and mutual respect and that they treat every human being equal. So who are superior human beings? Find your own answer.

The fact is, if you are a humanist, you are not alone—about 1.5 billion people in the world today do not associate with any religion, which is almost one-fifth of the world's population. So I invite you to join me and be a humanist, because you won't be alone. Let us assume that among nonreligious people, half of them say they believe some sort of "spirit." There are still more than half a billion people globally who are either atheists or agnostics or rationalists or naturalists or cynics or freethinkers or deists or pantheists, such as spiritual, apathetic, nonreligious, or any irreligious description you can put on what is known as humanist. They are among the large group of people who are the third-largest life stances in the world after Christianity and Islam, according to Pew Research.

Let me give statistics of religious affiliation of Canadians. According to 2011 population census, highest proportion of Canadians is affiliated with Christianity, followed by the people who reported that they do not affiliate with any religion—meaning, they are the freethinkers, outside of the religious box people. I am pretty sure that 99 percent of Canadians don't know it. As you read these words from this book, my friends and relatives of faith may reject not only these words but me as well. More to it, some members of radicalized faith groups may have their own way of punishing me.

I would like to ask a simple question to the believers: What is your "prime" identity in the natural world? The most probable answers, I presume, would be "I am a Muslim," "I am a Christian," "I am a teacher," "I am Jewish," "I am a Buddhist," I am an imam," "I am a Sikh," "I am a woman," "I am a professor," "I am a man," "I am a rabbi," "I am a Hindu," "I am a human being," "I am an American," "I am a Canadian," "I am a philosopher," "I am British," "I am an Indian," "I am a scientist," "I am Chinese," "I am Japanese," "I am an Arab," "I am a priest," "I am Russian," "I am a doctor," and so on. What is your answer? If you have rightly identified the prime one (who you are), then identify the second and the third. Every human being is different according to their depth of knowledge, of experience, of teachings and also by culture and by religion and, of course, on their views of life. The answers will likely demonstrate the thinking process of the believers. Einstein said, "Two things are infinite—the universe and human stupidity. I am not sure about the universe." While believers will justify their position by their belief, humanists will justify their position by reason.

I doubt that all readers will find my presentation of the arguments to be excellent, but, rather, many may find them to be insufficient and unbalanced. I take responsibility for all the flaws in my presentation. At the same time, I must confess that as a new author, I put best effort to keep the arguments in order, probably with no full success. Moreover, it is likely that my understandings on most of the themes are subject to criticism by many who are the experts on specific subjects or by others who have commonsense knowledge. I am neither a scientist nor a philosopher but only a freethinker who chooses to take on this complex and controversial topic and to write this book. It is, of course, a very brave and daring attempt indeed. I would appeal to readers to take all these into consideration and get the message that I have intended to share with you all.

INTRODUCTION

I seek in vain to find a resting place; I trudge despairingly this endless road; How many thousands ages must we wait till hope springs blooming from the dusty earth?
—Omar Khayyam

Humans have taken on to measure things and gather knowledge of what is and what is not, what is known and what is unknown. Human beings are also said to possess divine intelligence that transcends the physical entity. They are seen as the inventors and discoverers. They are complex beings. Naturally, mankind is endowed with inquisitive instinct. This natural endowment has made them always ask questions and attempt to provide answers to their unending questions. The quest to understand and completely comprehend the mysteries surrounding them has led mankind to invent several epistemological approaches. These are aimed at actualizing themselves in the world they find them, which they cannot adequately explain. This desire for rational explanation of the complex world and the many unknowns has led men into the discovery of philosophy, science, and religion as paradigms of explanation. Along with these, they developed other ideas and concepts such as humanism, existentialism, pragmatism, and rationalism, among others. Of all these, humanism seems to be the most challenging, as it has elements that could make it philosophical, scientific, and religious. This is more varied, as there are branches such as literal humanism, Renaissance humanism, cultural humanism, philosophical humanism, religious humanism, Christian humanism, modern humanism, scientific humanism, secular humanism, and many others.[1] Admittedly, seven billion people on earth believe in humanity yet identified into or associated with different religious groups. This suggests that every religion promotes (it could be religiopolitical; it's debatable, though) humanity. If the key theme of religion is humanity, then the sum of all religions is a religion that can be called humanism. What is humanism? How does it encompass science, philosophy, and religion? How can it be practiced in the twenty-first century? There are more questions than answers.

Our planet revolves around a medium-sized star (the sun), which is situated near the edge of a galaxy of approximately three hundred billion stars, an integral part of a group of galaxies consisting of several other galaxies, which is part of an ever-expanding universe that also contains one hundred or more billion galaxies. Our species have been in existence since only a very short time on planet Earth. It is also revealed that Earth itself has been in existence since only a short time in the history of our galaxy. So our existence as a species is, therefore, incredibly minuscule and is a brief part of a much larger picture. In the light of this cosmetic reality, is it not very curious to find that (in the absence of any direct and tangible evidence) religious thinkers and believers can conclude that the universe or some superpower beyond it is concerned with and mastering our well-being or our future? From all these cosmic explanations, is it not logical to think that we alone are concerned and responsible for our own well-being, future, and destiny?

We, as human beings, are part and parcel of the natural world. We marvel at the vastness of the cosmos and are content in relishing our place within it. We experience the universe by great fascination and inspiration because we're made of the same ingredients as butterflies and blue whales, giant banyan trees and spiral galaxies. The most beautiful gift we have had is life on Earth, a part of the universe. Life comes once, and it is very short. So it is important that we enjoy our lives to the fullest we can within our ability and power. In the midst of everything in our lives, cry of aching pains and anguish gnaws the heart. But those who lose freedom of choice and speech, absorb defeat from human dignity, and distaste life's beautiful gifts are bound to be foolish. Life without open rational thinking and not doing excellent in human endeavor is slavery—the conditions in life are synonymous to thinking and walking all over within an invisible fence or even on barbed wire and indulging in a thinking process in confinement into a closed box.

We put emphasis on the scientific inquiry, which enables us to construct knowledge and test their reliability and validity. This, in turn, emphasizes the universality of the scientific methods to find out the cause and effect in the physical world at both the organic and inorganic levels of existence, and so it is in human behavior because humans are part of nature. Another aspect of human distinctiveness is the capacity for moral values; we may call it morality. We are capable of making good choices, which, in turn, function to develop and shape individual characters. We can function collectively with moral values that are necessary for universally applicable. These are universal human values we are capable of establishing in our time without hovering around various religious doctrines and superstitions.

We must see one another as fellow human beings—equal in dignity, respect, and compassion in our pursuit of life. We deserve to be cherished by our loved ones and treasure our short lifetime to face challenges and tackle problems using the most successful methods such as science, reason, free inquiry, and seeking the truth to improve our lives and enhance the well-being of every one of us. We deserve not to fear the unknown and unexplained metaphor but, rather, to take courage from the wondrous discoveries that have been made so far. We accept that our lives will end, yet we find hope and take great joy in knowing that life goes on. We see ourselves as one tale among millions in the magnificent and ever-evolving story of life, and we are thrilled to be a part of the natural world. We deserve to be rational secular human beings practicing humanity and living outside the religious box. It is a testament that more than a billion people on earth feel that religion is increasingly unnecessary, unimportant, and irrelevant to life in the twenty-first century and onward and that we should assert positive human values free from superstition, religious dogmas, and blindness by innovating thinking and adopting change. I quote someone who said, "I am fortunate to be a part of the human experience. I am the beneficiary of past and present intellectual giants whose compassion and creativity set the foundation for a brighter future. As a part of this human experience, I joyfully carry that torch with the same love and compassion that my friends, loved ones, and complete strangers afford me every day, because progress rests in our collective hands."[2]

Normally, most people don't choose a religion; rather it is simply drummed and imposed into their brains at an early age. Think awhile. How can we explain that a Catholic country produces millions of Catholics rather than Muslims and vice versa? Well, don't get offended by what I'm saying, but, rather, put yourself outside your own head, the box, for a moment and think about it. Have you ever seriously thought to prove the validity of your own beliefs, or did you just acquire them by osmosis, gradually absorbing ideas without questioning? It is true that the religious scholars, intellectuals, rationalist thinkers, and even the majority of the educated religious people with analytic minds could not, in the past and even today, denounce and reject religion altogether as it was ingrained in their subconscious minds. This is the most powerful narcotic if it is injected to a person from childhood (normally, it happens this way). This is exactly the brain chemistry that works for every religious person, except a few who are openly critical about the holy books and the so-called prophets and saints. The uneducated and not fully rational common populations believe and practice the rituals without thinking and knowing what they believe in and practice. On the contrary, for some of the intellectuals and rational people, it may be difficult to accept the holy books at face value, and thus, they try to justify esoteric meanings in the verses of the scriptures and in the words of the prophets and saints. I caution, however, not to misinterpret what I am trying to make clear. My purpose is not to disrespect what religion you believe in and practice. Rather, I'm just pointing out that such beliefs (if not chosen rationally but absorbed ideas from a very early childhood to a growing-up age; mostly they are) can be counterproductive to your chosen life.

Religion has been one of the great forces that shaped the human history and transformed the life of the people on earth. Religion is extremely complex, intricate, variegated, and full of paradoxes. It includes thousands of rituals like celebration, despair, prayer, ethical vigor, mystery, retreat, social activism, monastic quietude, contemplation, animal sacrifice, incurring physical pain, torture, demonstration of terror, images of hope, symbols of fear, affirmation of life and struggle against death, superstitions, many beliefs and dogmas about natural and supernatural things, and so on. For centuries, people attempted to know, why religion? Also, there's the fact that there is a widespread element in our society that religious faith is vulnerable to offense and must be protected at any cost by strong and thick walls as if it is fragile, like glass that needs to be protected by a steel fence. Religion is in the heart of theists, and it so sacred or holy or whatever. It is a notion that no one is allowed to say anything bad or skeptical about it. One can be skeptical about anything but not religion. Everyone with some religious faith gets furious and frantic about it because no one is allowed to say anything bad about religion. The fact is that religion is a demanding subject for criticism, particularly in a pluralistic society, and that the stigma associated with nonreligiousness has to be removed to restore human development.

For religion, the reasonably accepted explanation is that primitive humans found themselves in a dangerous and hostile world and encountered constant fear of animals, of not being able to find enough food, of injury or disease, and of natural phenomena like thunder, lighting, volcanoes, flood, etc., which were constantly appearing in their lifetimes. Finding no security, they created different gods to give them comfort and courage to face calamities of life. Even today, people ask for God's help at the time of crisis and become religious. They say that the belief in a god or gods give them the strength and courage they need to deal with life's adversities. All these support the doctrine that the god idea is a response to fear, frustration, insecurity, crisis, and all other adversities of life. As a superior species, human beings do not like chronic anxiety. Thus, some coping mechanism had developed inception. So our ancestors conceptualized animism (meaning breath or soul), which was the first bona fide religion whose roots go back to the Paleolithic period.* In this belief system, a soul or sprit existed in every object, even if it was inanimate. The spirit was thought to be universal and came to be god. A belief in God is found in all societies worldwide and throughout history. Some scientists believe that this capacity may be designed into the brain's circuitry through evolution to facilitate altruism and cooperation among individuals and bring order and stability to society.

Michael Shermer (a Christian who became an atheist), who has done extensive research on people's belief in God, concluded that most of the time, people believe God or some weird things because of a variety of reasons without empirical evidence and logic. Instead, factors such as genetic predispositions, parental predictions, sibling influences, peer pressures, educational experiences, life impressions, etc., are all that shape the personality preferences and emotional inclinations, in conjunction with numerous social and cultural influences, and lead them to certain beliefs. It is not the case that anyone sits down and studies a bunch of facts and weighs the pros and cons and chooses the most logical and rational one. Rather, they gather the facts of the world that come to them through the colored filters, ideologies, hunches, hearsays, religious declarations, and prejudices that have accumulated throughout their lifetime. Then they sort out the information and select those that confirm what they already

---

* Paleolithic period is the Old Stone Age, the earliest period of human development and the longest phase of mankind's history. It is approximately coextensive with the Pleistocene geologic epoch, beginning about two million years ago and ending in various places between forty thousand and ten thousand years ago, when it was succeeded by the Mesolithic period. By far the most outstanding feature of the Paleolithic period was the evolution of the human species from an apelike creature, or near human, to true *Homo sapiens*.

believe and ignore those that disconfirm. Most of us do this.[3] Many other psychologists have similar views. For example, Gary Marcus said that we always seek to confirm what we believe and that we make up reasons to believe them even by disconfirming evidence.[4] Social psychologist Carol Tavris said, "Most people, when directly confronted with proof that they are wrong, do not change their point of view or course of action but justify it even more tenaciously."[5] But there are people who are better critical thinkers than others, and they are smart to be skeptical about the faith and religion they belong to. When they think about their religion and cannot reconcile it with the things they have learned and experienced in life, they may reject the religion or faith.

Drawing two parallel lines serves as a metaphor to define two distinct ideologies human beings follow as the supreme path for life—religion and humanism. Humanists do not defend religious superstitions, nor do they defend religion's violations against humanity, thus human rights and dignity. Yet many people try to understand why a vast majority of people are religious. Of course, one has the right to practice the religion of one's choice. But it is also the basic right and freedom for anyone to criticize religion as need be. Religious people in most of the religions, in many instances, are most extreme and violent.

Humanity evolved in the early stages of rapid expansion in knowledge, freedom, intelligence, and wisdom. Humanism, as such, is the progressive philosophy of life that, without superstition and spirituality, affirms our ability and responsibility to lead ethical lives of personal fulfillment that aspire to the greater good of humanity when all the attributes of human quality are appropriately used. It is such a simple, elegant statement that nobody could argue with. According to the *Humanist Manifesto*, wonder, imagination, fulfillment, creativity, and meaning are available to everyone, no matter religious or not. These emotional states are parts of our human birthright. Our fullest possible development of our lives with a deep sense of purpose, finding wonder and awe in the joys and beauties of our existence, in challenges and tragedies, and even in the inevitability of death all lead but to the fulfillment of life. Humanism relies on the rich heritage of human culture, which provides comfort in times of sorrows, pains, and despair and encouragement in times of thriving, achievement, and progress. Humanism promotes a world where violence and fear are not the means to achieve ideals and goals. In every case forms, extremism must be condemned. Neither should fear nor ignorance be permitted to sanction prejudice, discrimination, racism, and humiliation. Humanism suggests that religious liberty is the freedom for all—freedom to peacefully affirm and practice a faith, freedom from religious coercion, and freedom to peacefully reject a religious faith. This is what religious liberty should stand for. This is the central tenet of humanism. Bertrand Russell said, "There is no reason to believe any of the dogmas of traditional theology and, further, that there is no reason to wish that they were true. Man is free to work out his own destiny. The responsibility is his, and so is the opportunity."[6]

In a mostly religious world where much of society still believes in heavenly afterlives, some people are skeptical about whether extending human life spans is philosophically, morally, and scientifically correct. But religions oppose most forms of progress. For example, religions oppose the use of birth control, women's and civil rights, stem cell research, genetic engineering, and scientific advancement in general. Religion is from our past; it opposes the future. Can we function without the religious beliefs? Humanists think we can. We just need new principles based on scientific worldview. Such new principles come from the idea of transhumanism, a new philosophy that has been proposed to continue the ideas of humanism in a new world where science and technology are the major drivers of change. Whatever shape those ideas take, they suggest that humans can evolve into something much more than they are now.

Transhumanism is a cultural and intellectual movement that promotes that we can improve the human condition through the use of advanced technologies. Max More defined transhumanism as "philosophies of life that seek the continuation and acceleration of the evolution of intelligent life beyond its currently human form and human limitations by means of science and technology, guided by life-promoting principles and values." He also emphasized that transhumanism is a dynamic interplay between humanity and the acceleration of technology.[7] Julian Huxley, the English evolutionary biologist and humanist, said that the human species can transcend itself, not just sporadically (that is, an individual here in one way and an individual there in another way) but in its entirety as humanity. He suggested a name for this new belief as transhumanism, meaning man will remain man, but transcending himself by realizing new possibilities of and for his human nature.[8]

It is true that the future is unborn and impossible to predict. But that is not going to stop people to think and predict. Our social habits, our ways of thinking, and our invention of various technologies are the driving forces to move forward for better existence without losing the virtue of humanity. It is reasonable to think that our understanding about ourselves and about our relationships with nature has increased significantly because of the continuous advancement of science and technology. We see that the reality is not static, since humans, like the rest of nature, are dynamic and are changing constantly. Transhumanism, therefore, transforms static ideas of humanism because human beings evolve by themselves at a certain speed. Even in Darwin's perspective, it is conceivable that humans are not the end of evolution but that it is just the beginning of a conscious and technological evolution in modern time.

Transhuman technology allows us to upgrade ourselves in terms of our physical strength and power, our intelligence, our cognitive and emotional senses, our longevity, and any other physical and mental capacity. But transhumanism is not a distant matter that is very far away from happening. Rather, there are plenty of technologies that are available right now that are transhuman. For example, magnetic implants, implanting chips and electrodes, night vision drops, virtual reality and augmented reality, and many others. We, however, should not forget that transhumanism is not just about all these technologies; rather by adapting technology only, our lives are not complete. We need social, philosophical, and political establishments, disciple, and progress, along with the technological and biological advances, for transhumanism to become a reality.[9] The Bible said that God made man in his own image. The German philosopher Ludwig Feuerbach said that man made God in his own image. The transhumanists say that humanity will make itself into god.[10]

This book is about freethinking—no-nonsense, unemotional, nonirrational view of the natural world, the meaning of life, morality, and self-consciousness. The point is that the religious believers are very much aware of the arguments against religion. But they make clever excuses to justify their faith convictions to be true. Generally, common believers do not bother to think analytically about their faith; they just follow texts and credible people's sayings. So it is so much easier for the commoners to conceive the faith for granted. How can they get out of religious prison? How can they know the world outside religion? They can't. This book is for them. This book is for the swing people, those who are standing on the hanging balance between religion and nonreligion and also who are devoid of courage to cross over the platform from religion to humanism, from believing to thinking. This book is also for those who not only have developed doubts but also have already mentally denied the religion.

CHAPTER ONE

Religion

Fear is the main source of superstition and to conquer fear is the beginning of wisdom.
—Albert Einstein

Rational arguments don't usually work on religious people. Otherwise, there wouldn't be religious people.
—Doris Egan

Roots of Religion

Through human history, people across the world have asked questions about the world and people: Who are they, and what is their purpose in life? These questions have different answers for different people. Some believe that there is more to life than just physical body; they believe that prayer in some form or manner brings them closer to an ultimate power that exists beyond the everyday world. Many others have different opinions and argue that by living in a certain way in their lifetime, the answers can be found.

The word *religion* comes from the Latin word *religio*, meaning "a duty." Each and every religion compels their believers to strictly obey and follow a set of rules, guidelines, and rituals. On the contrary, in some society, there is no such word as *religion*; to them, religion is a part of their daily-life activities, and a special word for religion is not required. In fact, it is impossible to define religion in a way that everyone finds acceptable, as it does not fit to everyone's perspective of life and higher consciousness.

1

Another word for religion is *faith*. This originates from certain questions such as, Why was the world created? How should people live? Why is there so much suffering? What happens after death? Any of these questions can be tested by reason or can be proved by scientific evidence. Rather, one has to believe and have faith in them.

People who have attempted to find the root of religion find that early humans lived in social groups who saw religion as a force that held communities together for survival, because it gave all members shared rules for living and a shared way of understanding the environment and the world around them. Psychologists and social anthropologists look at the way religion helps in reducing fear, gives consolation and comfort, and perhaps satisfies our quench to understand why we exist, with the belief and the idea that some power beyond humans can be relied on. Believing in religion is a cognitive way of coping with the ruthless and harsh world we live in. But the truth is, there is no or little evidence that religious belief protects us from disease. Even if it does, the evidence is not strong. Maybe some faith healing turns out to work in some cases. But it would not be necessary to give credit to the true value of religion's claim, said Richard Dawkins (2006). He quoted George Bernard Shaw as saying, "The fact that a believer is happier than a skeptic one is no more to the point than the fact that a drunken man is happier than a sober one."

However, everyone does not believe in religion. Many think that religion blocks people's mind-set from being confident in what they can do for themselves by consciousness, intelligence, and knowledge. The people who don't believe in religion think that it is perfectly possible to lead a good, honest, and fulfilling life without religion. Those who don't believe any supernatural power are called atheists. Others think that it is impossible to know whether or not God exists, as there is no proof either way; they are known as agnostics. Many people are humanists; they think that human beings have the capacity to develop, function, and flourish within them and can build a prosperous, healthier, happier, and caring world. They also think that people deserve to be judged on their own merits and karma rather than by a rigid set of religious rules and guidelines. We will discuss more about these religious-free, open-minded people in later chapters.

Organized religion traces its roots back to the Neolithic revolution* that began eleven thousand years ago in the Near East but may have occurred independently in several other locations around the world. The invention of agriculture transformed many human societies from a hunter-gatherer lifestyle to a sedentary lifestyle. The consequences of the Neolithic revolution included a population explosion and acceleration in the pace of technological development. The transition from foraging bands to states and empires precipitated more specialized and developed forms of religion that reflected the new social and political environment. While bands and small tribes possess supernatural beliefs, these beliefs do not serve to justify a central authority, justify transfer of wealth, or maintain peace between unrelated individuals. Organized religion emerged as a means of providing social and economic stability through the following ways:

- Justifying the central authority. This in turn possessed the right to collect taxes in return for providing social and security services.
- Bands and tribes consist of a small number of related individuals. However, states and nations are composed of many thousands of unrelated individuals. Organized religion served to provide a bond among unrelated individuals who would otherwise be more prone to enmity.
- Religions that revolved around moralizing gods may have facilitated the rise of large cooperative groups of unrelated individuals.

The states born out of the Neolithic revolution, such as those of ancient Egypt and Mesopotamia, were theocracies, with chiefs, kings, and emperors playing dual roles of political and spiritual leaders. Anthropologists have found that virtually all state societies and chiefdoms from around the world have been found to justify political power through divine authority. This suggests that political authority co-opts collective religious belief to bolster itself.[11]

Since the dawn of human consciousness, men and women have regarded the supernatural world with a mixed sense of awe, fear, and hope. They sought to bring their lives into harmony with the supernatural world. Early religion had a strong basis in the sense that life was a struggle

---

* The Neolithic Revolution or Neolithic Demographic Transition, sometimes called the Agricultural Revolution, was the world's first historically verifiable revolution in agriculture. It was the wide-scale transition of many human cultures from a lifestyle of hunting and gathering to one of agriculture and settlement, which supported an increasingly large population.

for survival and a readiness for death. So men practiced sacrifices and prayers to the mysterious forces or spirits, which they believed to control the workings of nature with the hope that those forces and spirits would eliminate catastrophe, ensure good hunting, obtain better harvests, and again after death. Even prehistoric men seem to have shared, in all religious ideas, the belief that in some way, a person lives on after the death of their body. The evidence of ceremonial burials and the provision of food, utensils, and weapons for the use of the dead on their journey into the next life go back to the Neanderthal men who lived in Europe more than fifty thousand years ago.

With the rise of civilization in the lands of the Mediterranean and the Middle East (Egypt, Mesopotamia, Greece, and Rome in particular), the idea of "otherworld," of innumerable gods and goddesses, most of them depicted in recognizable human or animal shape, came into the minds of people. Each culture had its own cluster of divinities, usually ruled by super god. This multiplicity of gods (called polytheism after two Greek words meaning "many gods") was the religious idea in the ancient world. Polytheism's roots lay in primitive man's attempts to understand the elements of nature such as storms, floods, rivers, fires, droughts, the sun, and so on. They regarded them as powers to be worshipped because they affected their lives so directly.

From the religious experiment that was carried out in Egypt 1,400 years before the birth of Christ, we came to know that the pharaoh Akhenaton turned his back on the traditional gods (the greatest of whom was Amun) and established the worship of one god, the sun god Aton. However, after the death of Akhenaton, Egyptian religion reverted to the worship of many separate gods. This brief phase of monotheism (the worship of one god) came in the century before the Jewish leader Moses lived in Egypt. It is possible that he was influenced by this belief. Moses set out to convince the Hebrews that the god Jahweh had chosen them, as they were uniquely his people and should serve him only. It was their duty to live by his Ten Commandments. According to the Bible, God revealed the commandments to Moses on Mount Sinai during the Hebrews' journey out of Egypt to Palestine—the Promised Land. This uncompromising set of rules (with its prohibition of crimes, which set man against man, such as adultery, covetousness, and theft, and its insistence on the duties owed to God, to parents, and to other people) was originally framed for an obscure tribe in the desert. After establishing itself in Palestine, the national religion of the Jewish people made remarkable progress, giving birth to ideas that had a revolutionary impact on world religion, especially Christianity and Islam.[12]

Religious concepts are human concepts with a long list of standard characteristics that make them sensible to our ordinary ways of thinking. These concepts are related to the human brain. When the temporal cortical areas of the brain are targeted for stimulations, subjects often report dreamlike visions, often with mystical or religious content. This supports the notion that temporal circuits are central to religious experience. Modern science is beginning to understand the neurological mechanisms that give rise to the religious experiences of the believer.

Biological Roots of Religion

All human cultures appear to include faith in supernatural powers. This tradition has played an active and essential role in how our ancestors perceived and adapted to their environment. Humans have thrived on earth because of their ability to solve problems by way of cause and effect. This would indicate that human minds are basically pragmatic, and yet most human beings have held religious beliefs based on no empirical evidence whatsoever. A considerable part of human behavior is based on our biology. Through the process of natural selection, by interacting with social influences, we developed genetic predispositions to behave in ways that ensured our survival as a species. The evidence of prehistoric skull sizes and shapes, ancient objects, and the customs of primitive peoples indicate that the advantages of linguistic communication favored individuals with greater neurological capacity for verbal communication and that the culture and genetics coevolved to produce the modern human brain and thousands of human languages. This is a paradigm for the development of religion. As Professor Burkert said, "We may view religion, parallel to language, as a long-lived hybrid between cultural and biological tradition."[13]

Morton Hunt—the author of the article "The Biological Roots of Religion: Is Faith in Our Genes?"[14]—summed up the sociobiological theory of the roots of religion: "Genetically built into early human beings was a set of mental, emotional, and social needs that caused culture to develop in certain ways, including the development of various religions, and caused culture reciprocally to favor and select for evolution those human traits that provided sociocultural advantages to the individuals possessing them." Morton wrote the article by posing a fundamental question, why are unbelievers so different from the overwhelming majority of their fellow human beings? In response, he stated that throughout civilized history, a small minority never needed supernatural religious explanations or of the mysteries, tragedies, and glories of everyday life. These minorities can be brought in the light of Spinoza's belief in God, who is coexistential with the actual universe that is the god who is neither outside the universe nor above it but identical with it and with all natural laws. For Spinoza, God is nothing more or less than the total corpus of universal laws.

Perhaps current unbelievers (who disassociate with any religion) are the followers of Spinoza, sensitive to and in tune with the god who pervades the universe, who is the universe, who is identical with reality. Perhaps unbelievers do not reject the religious needs and impulses of the human race in realistic and humanistic terms but replace the fairy tales of conventional religions by the more intellectually demanding tales provided by modern science of natural laws and of the demonstrable, replicable evidence of cause-and-effect relationships. Perhaps for unbelievers, scientific humanism offers deeply satisfying answers to all those profound and troubling mysteries that religion purports to answer, and unbelievers are comfortable with those answers, although they are incomplete and, no matter how our knowledge increases, will remain so with new discoveries, always raising new and more complex questions about reality.[15]

Neurological Origins of Religious Belief

Religious belief and behavior affect the brain like habits, emotions, and memories do. It is possible to identify specific chemicals, genes, and clusters of neurons that cause belief like religiosity. Rutgers University evolutionary biologist Lionel Tiger[16] thinks that religion is made by the brain. He said that the root of religious belief is caused by a secretion called serotonin, which provides people with feelings of well-being when it floods the central nervous system.

Architect of the brain is also a factor for religious belief, as scientists suggested. Religiosity is spread out along a neural network composed of the frontal, parietal, and temporal lobes. For example, decreased parietal lobe activity has been found to be linked to some religious experiences, and the decision-making and social aspects of religion are the functions of the frontal lobes. The connection between epilepsy and religious visions is the function of the temporal lobes. Epileptic seizures and the brain chemistry leads some patients to a gradual personality change, which, in turn, leads them to mystical and religious thinking, said neurologist Oliver Sacks.

Studies also suggested that the architecture of the brain itself and the changes therein caused by neurological and mental disorders might be a neurobiological basis for altered spiritual and religious behavior.

Religious feeling might be induced through nondivine means by altering brain chemistry or structure. To test this, the cognitive neurologist Michael Persinger[17] of Laurentian University, Canada, built a "God helmet" fitted with electromagnetic solenoids intending to induce religious experience. In 2003, evolutionary biologist Richard Dawkins tested the contraption without experiencing religious conversion. He suggested that if religiosity operates in specific parts and chemicals of the brain, then its origins might be written in the blueprints of life, our genes. Geneticist Dean Hamer[18] at the NIH found God in a single gene—vesicular monoamine transporter 2 (VMAT2)—and identified this as the "God gene," a leading gene, among many others, written into our genetic code that predisposes people to religiosity. It seems that learning a society's religion is hardwired into humans through inherited genes. Scientists have identified the neurotransmitter serotonin (a network of neurons in the frontal, parietal, and temporal lobes) and the gene VMAT2 (as chemical, structural, and genetic origin points) that may be responsible for religiosity.[19]

Meme Theory of Religion

A meme is an idea or behavior that spreads from person to person within a society. The word originates from the Greek word *mimeme*, meaning "imitated thing." The term was coined by Richard Dawkins in his book *The Selfish Gene* in 1976 as the mental equivalent of a gene. Dawkins proposed the idea that social information can change and propagate through a culture in a way similar to genetic changes in a population of organisms—evolution by natural selection. The idea was subsequently introduced into finding the root of religion, which was named *memeplexes*, because they contain vast numbers of interacting memes. Examples of memes are tunes, ideas, catchphrases, clothes, fashions, and ways of making pots or of building arches. Just as genes propagate themselves in the gene pool by leaping from body to body via sperms or eggs, so do memes propagate in the meme pool by leaping from brain to brain via a process that, in the broad sense, can be called imitation. The meme is also a unit of information, formed in the brains through feelings and thoughts, which are transmitted by body language and facial cues as well as through language, both written and spoken (Dawkins 2006).

Craig A. James[20] outlined meme theory to an understanding of the evolution of religion. Even one may argue that meme theory has no scientific basis; James suggested that it can serve as a practical tool to help explain much of human civilization. He recalled Dawkins's proposal of memes as mental constructs that continue to mutate in response to the social environment, which is analogous to the function of genetic DNA mutation in response to the changing physical environment. Moreover, memes also refer to specific beliefs and attitudes that may start and end in a given person or propagate in the human population much as viruses and bacteria, no matter if it is harmful or helpful or benign, until they mutate or die out. For example, in the evolution of core religious beliefs from the time of the Abraham myth (circa 1900–1800 BCE) to the time of Jesus, it is strongly argued that some set of memes was the cause of the progression of religious beliefs and attitudes during barbaric and suppressive regimes of the Greeks, Romans, and Jews, because of which Christianity was born.[21]

The greatest proponent of memetics, since Dawkins, has been the philosopher Dan Dennett. In his work, which appeared in 2006, titled *Breaking the Spell: Religion as a Natural Phenomenon* (Viking, New York, 2006), Daniel C. Dennett explained his ideas on memes and the theory of memes by applying it to the study of religion from the perspective of evolutionary biology. His conclusions establish that religion is a meme and that its persistence in history is explained by the replicating processes of memetic structures. However, are there reasons of philosophical or scientific rationality for men having persisted in religion? Dennett does not go into a deep rational analysis of religion. He simply states that it has a memetic structure, and he considers that this is a sufficient basis to break the spell. According to him, religion is a conduct produced by a memetic structure that is produced by human psychism and is transmitted by memetic traditions registered in the memories of persons. This enables religion to be replicated and reproduced, which persists throughout generations. In his book, he describes that religion grew in primitive times because of irrational fear and threats of nature's calamities. This was constituted as a meme that has been unceasingly replicated down to us.

The analysis of meme theory of religion involves complex topics, ranging from quantum mechanics to psychology, philosophy, the theory of mind, and cosmology. Undoubtedly, within these spectrums, the discussions about the universe, human life, and religion continue to be an enigma. To billions of the world's population, nonreligion (such as atheism, agnosticism, and humanism) is possible, as is religion to many. Both have arguments that must be evaluated from the unbiased rational freedom of mankind. Therefore, one can argue that our positions could be atheism, agnosticism, and humanism is always a possible and respectable option, as are all the manifestations of human freedom. So there is no room to demonize or to humiliate nonbelievers of any religion for their point of view that religion is not essential to become good, moral, and fair human beings.

Psychological Theories of Religion

The main focus of the psychological approach to religion is the way in which religion operates in the mind of the individual. Sigmund Freud is the key scholar in this field, who gave explanations of the genesis of religion in his various writings. According to Freud, believers shape their personal image of God through a process of identification with parents (objects) during infancy, from which parental images are formed. God stems from the adult's projection onto the supernatural the dependency needs of the infant that are met by the infant's caregiver. In other words, God takes the place of the mother. Let us see how it works.

The child comes into the world without a fully developed brain. The interactions the child has with the world, especially with her caregiver (usually the mother), helps form the mind/brain. Specifically, the caregiver is taken inside the emerging mind and becomes an internalized presence inseparable from the developing self. The early interaction is imprinted on the brain. The infant has learned that when she cries for help, she can get from her caregiver whatever is needed. The mind/brain has been formed to believe that there will always be someone there to take care of. This expectation does not cease when we grow up, because it has already been ingrained into the brain synoptically. Brain research has also shown that our minds have not developed the ability to remember events or interactions during the first two to three years of life. Imaging studies have shown that the structures necessary to form memories are not functioning in infants' minds. Therefore, we do not explicitly remember the early dependence on and interactions with our caregiver. What we do remember is that our needs were met by another, and we remember the feelings of satisfaction and contentment. But this memory stays in our unconscious. Freud referred to this failure of memory as "infantile amnesia," and neurobiology has proved it.

We remember our early experience implicitly when we experience a physical or emotional crisis and feel the need for help or when we desire something special. Our experience as an infant has taught us that by asking for help, we can get it. When we no longer receive that help from an all-powerful parent, we must seek that help from another source, a higher power. By this time, we have been taught to believe in the existence of a supernatural caregiver who loves and protects. So we transfer our cry for help to him. Faber writes, "We turn to God as a child in distress turns to its mother." Another way of stating this is to say that when we experience a crisis, our mind/brain revisits and reuses the neural pathways it developed when we were infants in our parent's care. However, since we have separated from the parent and since we now believe in a supernatural parent, we cry out to that parent replacement. Because of infantile amnesia, we do not make the connection between our two experiences or our two caregivers (Murry 2009).

Religions are systems that provide parameters within which people interpret and try to influence the world and attempt to explain the inexplicable. Religious beliefs are seen as particular ways of interpreting the world, different from everyday thoughts and actions. Natural selection has produced the human brain that ensured the survival and reproduction. Brains and psychological mechanisms coevolved in the process of providing solutions to adaptive problems. Religion is a natural phenomenon that results from the evolution of various cognitive systems. It is thus a by-product of evolved cognitive adaptations. In an evolutionary framework, the link between religiousness, language, and cognitive strategies is explained by focusing on the kinds of information-processing problems the human mind evolved to solve to survive and reproduce. Languages developed to facilitate sharing of information among individuals and leads to the ability to formulate more and more complex questions and answers about the world, about life, and, finally, about the ultimate questions regarding the origin and purpose of life.

Whenever people make God the object of thought, an intentional object is created. The intentional object is treated as an agent with beliefs and desires; thus, anthropomorphism occurs. Religious beliefs and interpretations, however, are shown to minimally and systematically violate expectations about folk physics (a person who is invisible), folk biology (a being who lives forever), and folk psychology (that is, a being who knows exactly what you think); therefore, religious beliefs are described as counterintuitive. Human mind evolved various complex and interlinking cognitive strategies that include language through which ideas about God are shared. Most people do not question the socially constructed religious ideas of God.

The role of religiousness as psychological mechanisms to regulate interpersonal relationship and intragroup cooperation to ensure the successful reproduction of genes is the essence of the social functionalist view. During the evolution of *Homo sapiens*, individuals who formed groups learned to cooperate to fight danger and stuck together. This adaptation ability as social phenomenon was the by-product of religion, and its purpose was to improve cooperation within human groups. Generally, a verbal promise of cooperation is not necessarily reliable, and therefore, a promise is not automatically worth much, but if a person demonstrates commitment through acts, that promise is viewed as more reliable. Religion as a form of communication by means of shared acts and rituals contributes to the attribution of trustworthiness of the believers among themselves. The enhanced trust serves to facilitate cooperation. Religiously prescribed behaviors, badges, and bans are costly means of communicating commitment to the beliefs, ideals, and values in a particular religious community, and they are strengthening coalitions. Participation in rituals that involve all the senses (singing, rhythmic movement, consuming specific food and drink like bread and wine) emotionally validates and cements the commitment to a shared belief in one or more supernatural agents who have full access to all information.

Altruism among kin is an evolutionary gift as it secures successful transmission and survival of genes. Religious communities function as networks of kin that provide major benefits to believers, for example, resources (money, time, and talents) that are shared to the mutual benefit of all in the congregation. The evolutionary theories share the common assumption that there are no distinctive religious motives but, rather, simply, religious means for satisfying nonreligious human needs such as survival and procreation—strong linkages hardwired into humans as by-products of evolutionary adaptations. Principles and questions that guide the scientific inquiry do not provide answers to metaphysical questions of meaning, which belong to another order and are reflected in the realities based on culture. Positivistic science, for example, asks questions such as how and why humans developed to be religious but cannot provide answers to questions such as the reason for the existence of the universe and what the purpose of being human is and what the meaning of creation is. All people across the spectrum of premodern, modern, and postmodern eras search for answers to the question of meaning. How and what is believed varies according to the cultural and religious traditions of the group.

People want to understand themselves and the world in which they live. This quest for understanding and meaning is ultimately a spiritual endeavor, as it entails searching for meaning beyond the self, thus transcending the self. All people live in a personal world within which they create their individual understandings, explanations, and definitions of life and the universe, which, together, form their worldview. A worldview includes time, place, and culture and consists of ideas taken from ordinary people, experts, social institutions, and personal experience. Cosmologic assumptions are ideas about the origin and nature of the universe and are informed by various religious creation myths. Ontological assumptions are notions about the nature of existence and include beliefs regarding God. Epistemological assumptions relate to the nature of knowledge of truth and how it can be known. These fundamental assumptions form the basis of a person's understanding of the world and life in general without asking questions they have in their mind. Religious beliefs emanate from notions about God or gods—some superhuman characteristics such as omnipotent, all-knowing, and able to function outside natural laws to do miracles. In epistemological terms, religious knowledge is mostly wrapped in metaphoric language and grounded in religious texts, rituals, songs, and narratives and can neither be confirmed nor disconfirmed empirically. Beliefs are principles of action; they strongly guide people's behaviors, emotions, and actions and even command them to die or kill others for religion (Harris 2006).

Modernity is not looked in the same way in all societies because societies are vastly different from one another according to their history, their way of living, and their understandings of the world and cosmology. We experience a variety of worldviews in the twenty-first century. We understand, choose, adopt, and give meaning to things according to current worldviews. Religion establishments and narratives are not defensible by science, but still they are not discarded by the believers. The reason is that religions serve many psychological functions of the believers.[22]

Understanding of Religion

Religion is man-made and has nothing to do with spirituality. Belief in God does not require a religion. Spirituality is native to the human being, and we are constantly in contact with that, which we cannot measure in physical terms. The belief in a higher power or god does not require religion. Religion is a man-made institution developed specifically for the control of mass people by a smaller group of people. The intermingling of spirituality and religion is a major problem. There can be no consensus on human morality and ethics until it is done without religious interference. I am willing to believe in a creator (although no knowledge has gone so far to define it) but not necessarily the being described in the Bible or Quran or any other religious text. One may argue that the degree of order and the complexity of that order may demand the existence of a creator, and the observable physical laws are obvious, but there may be spiritual laws that underpin reality as well. Those laws are the foundation of ethics and morality. Religious people believe that religion is the tool by which the connection or the rules of restricting behavior are accomplished. But it is not the case because the inherent bias that each religion has interferes with the process. The control of the perception of truth influences the teaching so badly that most people only adhere to those rules and principles out of fear.

Robert Green Ingersoll delivered his last public address before the American Free Religious Association, Boston, on June 2, 1899.[23] He mentioned that for many centuries, it was believed that God demanded sacrifices, that he was pleased when parents shed the blood of their babes, that God was satisfied with the blood of oxen, lambs, and doves, and that in exchange, this god gave rain, sunshine, and harvest. It was also believed that if the sacrifices were not made, this god sent pestilence, famine, flood, and earthquake. During all these years and by all religious peoples, it was believed that this god heard and answered prayers and that he forgave sins and saved the souls of true believers. Now the questions are whether religion was founded on any known fact, whether such a being as God exists, whether any prayer was ever answered, and whether any sacrifice of babe or ox secured the favor of this unseen god. Did an infinite god create the children of men? Why did he create the intellectually inferior? Why did he create the deformed and helpless? Why did he create the criminal, the idiotic, and the insane?

If this god exists, how do we know that he is good? How can we prove that he is merciful, that he cares for the children of men? If this god exists, he has, on many occasions, seen millions of his poor children plowing the fields, sowing, and planting the grain, and when he saw them, he knew that they depended on the expected crop for life, and yet this good god, this merciful being, withheld the rain. He caused the sun to rise, to steal all moisture from the land but gave no rain. He saw the seeds that man had planted wither and perish, but he sent no rain. He saw the people look with sad eyes upon the barren earth, and he sent no rain. Do we prove that this god is good because he sends the cyclone that wrecks villages and covers the fields with the mangled bodies of fathers, mothers, and babies? Do we prove his goodness by showing that he has opened the earth and swallowed thousands of his helpless children or that, with the volcanoes, he has overwhelmed them with rivers of fire? Can we infer the goodness of God from these facts?

The power works for righteousness, but what is this power? It is the accumulated experiences of the natural world by which we get power and force for righteousness. For example, a child charmed by the beauty of the flame grasps it with the hand. The hand is burned, and after that, the child keeps its hand out of the fire. The power that works for righteousness has taught the child a lesson. This power and force are not conscious, not intelligent. It has no will, no purpose. It is a result. Religious people have tried to establish the existence of God in relation to morality and conscious. They insisted that the moral sense—the sense of duty, obligation—was imported, not produced by men but from God it comes.

Man is a social being. We live together in families, tribes, and nations. The members of a family, of a tribe, of a nation, who increase the happiness of the family, of the tribe, or of the nation, are considered good members. They are praised, admired, and respected. They are regarded as good (that is to say, as moral). The members who add to the misery of the family, the tribe, or the nation are considered bad members. They are blamed, despised, and punished. They are regarded as immoral. The family, the tribe, or the nation creates a standard of conduct, of morality. Conscience is born of love, and the sense of obligation, of duty, was naturally produced. There is nothing supernatural in this. Among savages, the immediate consequences of actions were the only motive. As people advance, the remote consequences are perceived. The standard of conduct becomes higher. The imagination is cultivated. A man puts himself in the place of another. The sense of duty becomes stronger, more imperative. Man judges himself. He loves, and love is the commencement, the foundation of the highest virtues. He injures one that he loves, and later comes the regret, repentance, sorrow, and conscience. In all this, there is nothing supernatural.

Religion has never made man merciful, has never made man free, and has never made man moral and honest. To those who believe in equality, uniformity, and fair justice among people of different race, color, ethnicity, and gender, religion is not the answer. In terms of the natural world, we cannot affect the qualities of substance and natural phenomena by prayer and worship. We cannot change winds and tides by sacrifice. We cannot cure disease by supplication. We cannot add to our knowledge by religious ceremonies. What is religion? Religion rests on the idea that nature has a master and that this master will listen to prayer and worship, that this master punishes and rewards, and that he loves praises and flattery and hates the brave and free.

For thousands of years, men and women have been trying to reform the world. They have created gods and devils, heavens and hells. They have written sacred books, performed miracles, built cathedrals and dungeons. They have crowned and uncrowned kings and queens. They have tortured and imprisoned, flayed alive and burned. They have preached and prayed. They have tried promises and threats. They have coaxed and persuaded. They have preached and taught and, in countless ways, have endeavored to make people honest, temperate, industrious, and virtuous. They have done their very best to make mankind better and happier and have not succeeded. But why?

The world is populated by the people's ignorance, poverty, and vice. They are not intelligent enough to think about consequences of their actions or to feel responsibility. People do have the attributes of vice, envy, jealousy, greed, and hate with limited love and affection. Majority becomes vicious; they live by fraud and imposture, and they pass on their vices to their children. Nature has no intelligence, no purpose. It sustains without intention and destroys without thought. Humans have little intelligence, and intelligence is the only lever capable of raising mankind. They should be more intelligent, more conscious, and more driven by reason than be more passionate and driven by impulse.

We cannot reform people with tracts and talk. We cannot reform people with preach and creed. Passion has always been deaf. These weapons of reform are substantially useless. Criminals, tramps, beggars, and failures are increasing every day. Ignorance, poverty, and vice must be stopped. This cannot be done by morale persuade, by talk or by example, by religion or by law, by priest or by hangman, or by force. But there is one way it can be done, which is science. In contrast, religion can never reform mankind because religion is slavery. It is far better to be free, to come out of the religion's box, to leave the forts and barricades of fear, and to stand erect and face the realities of life and challenges of the future with courage. It is far better to forget all gods, their promises and threats; to feel within your veins life's joyous stream and hear the martial music, the rhythmic beating of your fearless heart; and to elevate your inner sense to do all useful things, to reach with thought and deed the ideal in your brain, to look with deep and steady eyes for facts, to find the subtle threads that join the distant past with the present, to take burdens from the weak, to defend the right, to increase knowledge, and to nurture and develop the brain.[24] This is real religion. This is real worship. This is the thinking outside the box.

Albert Einstein's Perspective

According to Einstein, feeling and longing are the motive force behind all human endeavor and human creation. But what are the feelings and needs that have led us to religious thought and belief? He explained it by saying that with primitive man, it is the fear that evokes religious notions— that is, the fear of hunger, wild beasts, sickness, and death. Since the early stage of existence, as the knowledge of causal connections was poorly developed, the human mind created illusory beings. So they tried to secure the favor of these beings by carrying out actions and offering sacrifices, which were, according to the tradition, handed down from generation to generation. He also identified social impulses to be the other source of religious crystallization and explained the fact that parents and the leaders are mortal and likely to make mistakes. Therefore, the desire for guidance, love, and support prompted people to form the social or moral conception of God—the god of providence, who protects, disposes, rewards, and punishes. Einstein pointed out that the Jewish scriptures illustrate the development from the religion of fear to moral religion, a development continued in the New Testament. Although the perception from a religion of fear to moral religion is obviously a great step in people's lives, the truth of the fact is that all religions are a blend of both types in varying degrees.

Another stage of development is the anthropomorphic character of their conception of God. Einstein called it cosmic religious feeling. Individual existence impresses a man as a sort of prison, and he wants to experience the universe as a single significant whole. The cosmic religious feeling appeared at an early stage of development in some belief systems, such as Buddhism, Hinduism, and others. Many religious geniuses of all ages are believed to have this kind of religious feeling, who know no dogma and that there can be no church. In every age of history, we find men who were filled with this highest level of religious feeling and were regarded as atheists, sometimes also as saints. Men like Democritus, Francis of Assisi, and Spinoza are a few of them. In answering a fundamental question as to how cosmic religious feeling can be communicated from one person to another, when there is no definite notion of a god and no theology, Einstein brought the most important function of art and science, which has awakened this feeling.

With this view in mind, he addressed the relation of science to religion to a very different level from the usual one. "Looking at the matter historically, one is inclined to look upon science and religion as irreconcilable antagonists," he said. The man who is thoroughly convinced of the universal operation of the law of causation cannot accept the idea of a "being" that interferes in the course of events. Having taken this position, the man doesn't need to use the religion for fear and for social or moral purposes. A man's ethical behavior should be based effectually on sympathy, education, and social ties and needs; no religious basis is necessary. Man would be in a poor way if he had to be restrained by fear of punishment and hopes of reward after death, as Einstein explained. He maintained that the cosmic religious feeling is the strongest and noblest motive for scientific research. It is cosmic religious feeling that gives a man the strength to understand the rationality of the universe in disentangling the principles of celestial mechanics and the realities of life. "In this materialistic age, the serious and devoted scientific workers who spend several solitary years are the only profoundly religious people," he said.

An irreconcilable conflict between knowledge and belief that was begun during the last century is still prevailing today. The opinion grew among people of advanced mind that belief should be replaced by knowledge, because belief that does not itself rest on knowledge is superstition and, therefore, has to be rejected. According to this conception, the sole function of education is to open up the door to thinking and knowing. According to Einstein, a person who is religiously enlightened and who has liberated himself from the shackles of his selfish desires can be preoccupied with thoughts, feelings, and aspirations because of their superpersonal value. What is important is the force of this superpersonal value and the conviction regarding its overpowering meaningfulness, regardless of whether anything is done to unite this content with a divine being like Buddha and Spinoza as religious personalities. So a religious person is devout of those superpersonal objects and goals, which neither require nor is capable of rational foundation. In this perspective, there appears to be no conflict between religion and science. Science can only ascertain what is but not what should be, but beyond of its domain value, judgments of all kinds are necessary.

Religion, on the other hand, deals only with evaluations of human thought and action. Conflict arises when religious believers insist on the absolute truth of all statements written in the scriptures, which is unacceptable within the sphere of science. Although the realms of religion and science in them stand on different platforms, they exist between the two strong reciprocal relationships and dependencies. Though religion determines the goal, it has learned from science anyway in the broadest sense, because the means will contribute to the attainment of its goals. Science can only be created by those who are thoroughly imbued with the aspiration toward the truth and understanding. This source of feeling springs from the sphere of religion. Any genuine scientist would have his profound faith. This is what made Albert Einstein express that science without religion is lame, religion without science is blind.

Albert Einstein, however, pointed out that this assertion does not qualify with the concept of God. During the early period of mankind's spiritual evolution, human fantasy created gods in man's own image. Men altered the disposition of these gods in their own favor by means of magic and prayer. The idea of God in the religions taught at present is a transformation of that old concept of the gods. Its anthropomorphic character is shown by the fact that men appeal to the divine being in prayers and plead for the fulfillment of their wishes, he commented. According to him, one of his salient insights about God is as follows:

> *Omnipotent, just, and omnibeneficient personal God is able to accord solace, help, and guidance; If this being is omnipotent, then every occurrence, including every human action, every human thought, and every human feeling and aspiration is also his work; how is it possible to think of holding men responsible for their deed and thoughts before such an almighty Being? On giving out punishment and rewards, He would to a certain extent be passing judgment on Himself. How can this be combined with the goodness and righteousness ascribed to him?*

Are religion and science irreconcilable? Is it impossible to overcome contradiction between religion and science? Can religion be superseded by science? Einstein posed these questions and tried to answer them, because they raised enormous dispute and bitter fighting for centuries. Let us see how Einstein answered these questions in his own terms. Science is methodical thinking directed toward finding regulative connections among our sensual experiences. It leads to methodical action if definite goals are set up in advance. While it is true that science, to the extent of its grasp of causative connections, may reach important conclusions as to the compatibility and incompatibility of goals and evaluations, the independent and fundamental definitions regarding goals and values remain beyond science's reach.

Religion, on the other hand, deals with goals and evaluations and with the emotional foundation of human thinking and acting, as far as these are not predetermined by the inalterable hereditary disposition of the human species. Religion is concerned with people's attitude toward nature at large with the establishment of ideals for the individual and communal life and with mutual human relationship. These ideals of religion are attempted to be attained by exerting strong influence on tradition and through the development and promulgation of certain easily accessible thoughts and narratives (epics and myths), which are apt to influence evaluation and action along the lines of the accepted ideals. This mythical content of the religious traditions comes into conflict with science whenever this religious stock of ideas contains dogmatically fixed statements on subjects, which belong in the domain of science. Therefore, it is important for the preservation of true religion so that such conflicts can be avoided when they arise from subjects that are not essential for religion to pursue.

With regard to the actual living conditions of present-day civilized humanity, one is bound to experience a feeling of deep and painful disappointment at what one sees. When religion prescribes brotherly love in the relations among the individuals and groups, the actual spectacle more resembles a battlefield than an orchestra. Everywhere, in economic as well as in political life, the guiding principle is the striving for success at the expense of one's fellow men. This competitive spirit prevails everywhere, destroying all feelings of human fraternity and cooperation and love for productive and thoughtful work, rather springing from personal ambition and fear of rejection.

There are pessimists who hold that such a state of affairs is necessarily inherent in human nature; it is those who propound such views who are the enemies of true religion. This implies that religious teachings are utopian ideals and unsuited to afford guidance in human affairs. The studies of the social patterns in certain primitive cultures, however, have found that such a defeatist view is wholly unwarranted. Under the hardest living conditions, this tribe has apparently accomplished the difficult task of delivering its people from the scourge of competitive spirit and of fostering in it a temperate, cooperative conduct of life, free of external pressure and without any curtailment of happiness. The interpretation of religion as advanced implies a dependence of science on the religious attitude, a relation that, in our materialistic age, is overlooked. While it is true that scientific results are entirely independent from religious or moral considerations, those individuals to whom we owe the great creative achievements of science were all imbued with the truly religious conviction that this universe is something perfect and susceptible to the rational striving for knowledge. If this conviction had not been a strongly emotional one and if those searching for knowledge had not been inspired by Spinoza's *Amor Dei Intellectualis*, they would hardly have been capable of that untiring devotion that enables man to attain his greatest achievements, said Einstein.

Many great scientists are thought to be religious in some way, but by examining their beliefs deeply, it turns out to be wrong. Richard Dawkins has clarified this issue in his book *The God Delusion*. Let me address this in the case of Albert Einstein. Among many, confusion is caused by failure to understand the true Einstein's religion and supernatural religion. The confusion arises from his famous quote: "Science without religion is lame, religion without science is blind." Einstein also said, "I do not believe in a personal God . . . If something in me which can be called religious then it is unbounded admiration for the structure of the world so far as our science can reveal it." By religion, Einstein meant something entirely different from what is conventionally thought by many: "It was, of course, a lie what you read about my religious convictions, a lie which is being systematically repeated. I do not believe in a personal God and I have never denied this but have expressed it clearly. If something is in me which can be called religious then it is the unbounded admiration for the structure of the world so far as our science can reveal it."

Thinking deeply, it seems that Einstein did not contradict himself. One must have the intuitive knowledge and intelligence to understand Einstein's religion. Dawkins pointed out that by "religion," Einstein meant something entirely different from what is conventionally thought of by many. Here are more quotations from Einstein in order for readers to give flavor of his religion. "I am a deeply religious nonbeliever. This is a somewhat new kind of religion." "I have never imputed to Nature a purpose of a goal, or anything that could be understood as anthropomorphic." "What I see in Nature is a magnificent structure that we can comprehend only imperfectly, and that must fill a thinking person with a feeling of humility. This is a genuinely religious feeling that has nothing to do with mysticism." "The idea of a person God is quite alien to me and seems even naïve." "To sense that behind anything that can be experienced there is something mind cannot grasp and where beauty and sublimity reaches us indirectly and as a feeble reflection, this is religiousness. In this sense I am religious. In this sense too I am religious with the reservation that 'cannot grasp' does not have to mean 'forever ungraspable.' But I prefer not to call myself religious because it is misleading. It is destructively misleading because, for the vast majority of people, 'religion' implies supernatural."[25]

His quote "God does not play dice" must be translated as "Randomness does not lie at the heart of all things." Einstein used God in a metaphoric sense. Like him, many physicists, like Stephen Hawking, occasionally slip into the language of religious metaphor, Dawkins suggested. More specifically, the statement "God does not play dice" refers to the randomness of physical phenomena at very small scales (i.e., the field of quantum mechanics). Einstein didn't like the idea that there is no principle for the way tiny particles like electrons, neutrinos, photons behave, and that was the problem. He eventually didn't manage to unify his theory, as his theory of special relativity didn't work when it came to these tiny particles. So he used to say, "God does not play dice." But now, according to Stephen Hawking, we know that God does play dice and that sometimes he throws it someplace dark so that we can't see it! That does all have to do with the randomness of some physical principles in quantum mechanics! But in general terms, Einstein didn't like randomness of things at all! "God does not play dice." This represents Einstein's opinion on quantum mechanics, which relies heavily on probabilistic models of phenomena, which is in gross defiance of classical physics. He thought nothing in the universe happens by chance. This is summed up by the expression "God does not play dice."[26]

Spinoza's God

Benedictus de Spinoza was a social and metaphysical philosopher famous for the elaborate development of his monist philosophy, which has become known as Spinozism. He was named Baruch ("blessed" in Hebrew) Spinoza by his synagogue elders and known as Bento de Spinoza or Bento d'Espiñoza but afterward used the name Benedictus ("blessed" in Latin) de Spinoza. Spinoza was born in Amsterdam in 1632 into a family of Jewish emigrants fleeing persecution in Portugal. He was trained in Talmudic scholarship, but his views soon took unconventional directions that the Jewish community discouraged. Spinoza is best known for his *Ethics*, a monumental work that presents an ethical vision unfolding out of a monistic metaphysics in which God and nature are identified. God is no longer the transcendent creator of the universe who rules it via providence but nature itself, understood as an infinite, necessary, and fully deterministic system of which humans are part. Humans find happiness only through a rational understanding of this system and their place within it.

The first part of the book *Ethics* addresses the relationship between God and the universe. Tradition held that God exists outside of the universe, created it for a reason, and could have created a different universe if he so chose. Spinoza denies this point and said, "By God, I mean a being absolutely infinite, that is to say, a substance consisting of infinite attributes, each of which expresses eternal and infinite essence. God is infinite, necessarily existing, single substance of the universe. There is only one substance in the universe is God and everything is in God."

The second part of the *Ethics* focuses on the human mind and body. Spinoza attacks several Cartesian positions: (1) the mind and body are distinct substances that can affect one another; (2) we know our minds better than we know our bodies; (3) our senses may be trusted; and (4) despite being created by God, we can make mistakes—namely, when we affirm, of our own free will, an idea that is not clear and distinct. Spinoza denies each of Descartes's points. Regarding (1), Spinoza argues that the mind and the body are a single thing that is being thought of in two different ways. The whole of nature can be fully described in terms of thoughts or in terms of bodies. However, we cannot mix these two ways of describing things, as Descartes does, and say that the mind affects the body or vice versa. Moreover, the mind's self-knowledge is not fundamental: it cannot know its own thoughts better than it knows the ways in which its body is acted upon by other bodies. Further, there is no difference between contemplating an idea and thinking that it is true, and there is no freedom of the will at all. Sensory perception, which Spinoza calls knowledge of the first kind, is entirely inaccurate, since it reflects how our own bodies work more than how things really are. We can also have a kind of accurate knowledge called knowledge of the second kind, or reason. This encompasses knowledge of the features common to all things and includes principles of physics and geometry. We can also have knowledge of the third kind, or intuitive knowledge. This is a sort of knowledge that somehow relates particular things to the nature of God.

In the third part of the *Ethics*, Spinoza argues that all things, including human beings, strive to persevere in their being. This is usually taken to mean that things try to last for as long as they can. Spinoza explains how this striving "conatus" underlies our emotions (love, hate, joy, sadness, and so on). Our mind is, in certain cases, active and, in certain cases, passive. Insofar as it has adequate ideas, it is necessarily active, and insofar as it has inadequate ideas, it is necessarily passive.

The fourth part, "Of Human Bondage," analyzes human passions, which Spinoza saw as aspects of the mind that direct us outward to seek what gives pleasure and shun what gives pain. The "bondage" is domination by these passions or affects, as he called them. Spinoza considered how the affects, ungoverned, can torment people and make it impossible for mankind to live in harmony with one another.

The fifth part, "Of Human Freedom," argued that reason can govern the affects in the pursuit of virtue, which Spinoza called self-preservation: only with the aid of reason can humans distinguish the passions that truly aid virtue from those that are ultimately harmful. By reason, we can see things as they truly are, sub specie "under the aspect of eternity," and because Spinoza treats God and nature as indistinguishable, by knowing things as they are, we improve our knowledge of God. Seeing that all things are determined by nature to be as they are, we can achieve the rational tranquility that best promotes our happiness and liberates ourselves from being driven by our passions.[27]

Spinoza's most famous and provocative idea is that God is not the creator of the world but that the world is part of God. According to him, God is an absolutely infinite being—that is, a substance consisting of infinite attributes, each of which expresses eternal and infinite essence. Spinoza believed that everything that exists is God. However, he did not hold the converse view that God is no more than the sum of what exists. God had infinite qualities, of which we can perceive only two—thought and extension. Hence, God must also exist in dimensions far beyond those of the visible world. He derived an ethic from fundamental principles, which were closely linked to his view of "God or nature" as everything. The highest good, he asserted, was knowledge of God, which was capable of bringing freedom from tyranny by the passions, freedom from fear, resignation to destiny, and true blessedness. Spinoza's god is not the god of Abraham and Isaac, not a personal god at all, but his system provides no reason for the revelatory status of the Bible or the practice of Judaism or of any religion for that matter.[28]

He believed that anthropomorphic (humanlike) conceptions of God were childish and based on superstition, and he sought to replace such traditional religious beliefs with a more rational understanding of God and the world. In his *Theological-Political Treatise,* he will argue that miracles are not supernatural, the Torah was not written by Moses, and the prophets had vivid imaginations. Spinoza was critical of Judaism as well as Christianity. He is considered the father of what is known as biblical criticism. Spinoza believed the Torah as a human invention and the product of multiple people. He rejected the belief that the Jewish people are "chosen" and, therefore, superior to others, that prophets received revelation in the supernatural way, or that miracles were anything more than natural events.

There are comments about Spinoza's god as saying, for Spinoza, as for Einstein, God is imminent, not transcendent. This means that there is no god that transcends (exists outside of) nature. A transcendent god is kind of like a programmer god with nature as the software; this view usually takes humans as having a similar status to the programmer—they are both free. An imminent god inheres in everything, and everything must be seen as an expression of his existence. Notice how God here functions just like another symbol of ours—"reality." Reality inheres in everything, and everything must be seen as an expression of reality's existence. Everything follows by God's/reality's nature/laws just as the sum of the triangle's angles follow from its nature. The Abrahamic religions are incompatible with this worldview. The Genesis claims that God created nature for the sake of man and that man is free. For Spinoza, God equals nature. The idea that God watches out for us is an error.[29]

Spinoza was considered to be an atheist because he used the word *God* (Deus) to signify a concept that was different from that of traditional Judeo-Christian monotheism. Spinoza denied personality and consciousness to God. He has neither intelligence nor feeling nor will; he does not act according to purpose, but everything follows necessarily from his nature, according to law. Thus, Spinoza's cool, indifferent god differs from the concept of an anthropomorphic, fatherly god who cares about humanity.

There are similarities between Spinoza's philosophy and Eastern philosophical traditions. The nineteenth-century German Sanskritist

Theodor Goldstücker was the one who first noticed the similarities between Spinoza's religious conceptions and the Vedanta tradition of India. He wrote the following: "Spinoza's thought was a western system of philosophy which occupies a foremost rank amongst the philosophies of all nations and ages, and which is so exact a representation of the ideas of the Vedanta, that we might have suspected its founder to have borrowed the fundamental principles of his system from the Hindus, did his biography not satisfy us that he was wholly unacquainted with their doctrines . . . We mean the philosophy of Spinoza, a man whose very life is a picture of that moral purity and intellectual indifference to the transitory charms of this world, which is the constant longing of the true Vedanta philosopher . . . comparing the fundamental ideas of both we should have no difficulty in proving that, had Spinoza been a Hindu, his system would in all probability mark a last phase of the Vedanta philosophy. Striking similarities between Vedanta and the system of Spinoza—Brahman, as conceived in the Upanishads and defined by Sankar is clearly the same as Spinoza's 'Substantia.'"[30]

Spinoza's philosophy and ethics were scientific with the objective of attaining tranquility in the turbulent and conflicted world. His vision was not to escape from the world but, rather, to renovate it. He recommended changes in people's behavior, attitude, beliefs, and practices. He emphasized meditation on life but not on death. Spinoza was committed, throughout his life, to finding a way to unite science, ethics, and religion. In doing so, he promoted a metaphysical system that would make the whole nature, human life, and religious systems comprehensible. His works attempted to unify the nature, an ordered whole (Morgan 2006).

Today, he is considered one of the great rationalists of seventeenth-century philosophy, laying the groundwork for the eighteenth-century Enlightenment and modern biblical criticism. By virtue of his *magnum opus*, the posthumous *Ethics*, Spinoza is also considered one of Western philosophy's definitive ethicists. Albert Einstein named Spinoza as the philosopher who exerted the most influence on his worldview. Spinoza equated God (infinite substance) with nature, consistent with Einstein's belief in an impersonal deity. In 1929, Einstein was asked in a telegram whether he believed in God. Einstein replied by a telegram, "I believe in Spinoza's God who reveals himself in the orderly harmony of what exists, not in a God who concerns himself with the fates and actions of human beings."[31]

More Arguments

As humans, we have limitations in our endeavors and achievements. We can free ourselves from prejudices and blindness but not to the whole of it. As Michael Morgan has said, "Our goal is to free ourselves from the distortion and corruptions of our finitude to become rational, active, and free." This is something like to become the whole, which sounds like the highest good, or divine. This is a challenge in life that we should not escape. Life could be better from the status quo, and the religious doctrines, systems, and practices could be reformed, and the institutions could be renovated to serve human purposes.

Religions of the world and individual denominations or traditions within the religions teach very different beliefs about the existence of God, gods, the goddess, goddesses, etc. They have different views on the nature of deity, humanity, and the universe. But almost all share one belief, which is that they alone have the fullness of truth and that every other religion in the world is wrong. Even within a single religion, many denominations, traditions, and faith groups teach mutually exclusive beliefs, including the belief that they are right and all the other faith groups within their religion are wrong—at least to some degree. Religion is founded on local traditions and also by private revelation rather than any pragmatic evidence. There is a progression from primitive tribal animisms through polytheism to monotheisms (Dawkins 2006).

Twenty-first century is not the same as seventh century; things have significantly changed over time. But the followers of seventh-century religions believe that theirs is the sole truth. Is it the right way to move forward for better survival? I don't think so. I think that no religion should impose its own path and ideology to anyone and that no one is obliged to follow that path. It is not my intention to offend and hurt religious believers, but I am stunned by the disproportionate and unfair privilege and respect religions get compared to secular and freethinkers vis-à-vis those who do not associate with any religion.

Among the three Abrahamic religions, Judaism is the ancestor of the other two—Christianity and Islam. Judaism originally was a tribal cult of a single fiercely unpleasant god. During the Roman occupation of Palestine, Christianity was founded by Paul of Tarsus, a less ruthless sect of Judaism and a less exclusive one. Later, in the seventh century, Muhammad and his followers brought back the uncompromising monotheism of the original Judaism and founded Islam with a book called Quran. They introduced a very powerful method of military conquest to spread Islam. It is to be noted that Christianity was spread by the sword as well by the Romans after Emperor Constantine and then by the Crusaders and later by the *conquistadors* and other Europeans (that is, invaders, colonists) in the name of missionary accomplishment (ibid.).

The second-largest religion is Islam, and the believers are known as Muslim. They believe in the teachings of Muhammad. Their principal beliefs are that no god exists but Allah, each Muslim must pray five times a day, and they must fast during the month of Ramadan and make a trip to Mecca at least once during their lives. It is believed by Muslims that if these rules are followed, they would be alive and be united with Allah on the day of judgment. Muslims are mainly concentrated in the Middle East, northern Africa, and South Asia, including Indonesia and Malaysia, but also spread out in Europe and North America.

Jews make up another of the world's major Abrahamic religions, along with Christianity and Islam. Jews feel that they are the chosen people of Yahweh (God) because of the covenant that God made with the Jews through Abraham, who was considered the founder of Judaism and later through the Ten Commandments that were handed down from God to Moses. Jews think and claim the area around Jerusalem to be their sacred land because it was given to them by God thousands of years ago. Jews lived in this land called Israel for thousands of years until it was overrun by Muslims. This land in the Middle East is known again as Israel, where the main concentration of Jews lives. This was done so that Jews could have a homeland after all the persecutions of Jewish people in Europe during the WW II. Although people are somewhat familiar with the major religions, I haven't given emphasis to discuss them here, but I have presented a firsthand, brief description of each of them in the annexure.

The roles, relevance, and practices of world religion, at large, have become fascist. They have failed to offer for the pressing problems of humanity on the twenty-first century. Rather, religions have become the greatest liability for the entire humanity. All religions together spend more for war and defense industry than what they spend for poverty alleviation and human development.

Living in the fool's paradise, their advocates and propagandists claim that they are going to conquer the entire world by the false slogan of peace and harmony. In spite of their heavyweight and numerous prayers, sermons, sacrifices, celebrations, multibillion-dollar worth of worships, houses, and pilgrimages, and, of course, the empty promises of heaven or paradise, humanity is being trodden under the feet. In reality, though, religions have become the catalysts for long-term war, terror, poverty, inequality, and misery of the billions of people of the world (Mathew 2015).

Religion Wars and Challenges

War in the Religions

Religion divides people into separate groups and tells them that they are incompatible with one another. It provides justification for the killing of others and the promise of reward for martyrs. It justifies hate and murder. Religions are responsible for numerous conflicts and violence. Along with nationalism, religion is the major cause of war. Even many wars that were not directly caused by religion were used to justify it and motivate soldiers. For example, the fact that Protestant Germany invaded Catholic Belgium was a main part of the recruitment process in Ireland during World War I, even though it was not a purely religious war.[32] In recent past, the civil war between West Pakistan and East Pakistan, which killed three million people, gave birth to a new country—Bangladesh—in 1971. West Pakistan rulers used religion to escalate the war against the people of East Pakistan. Soldiers were told to kill East Pakistani people (Bengalis) as they become *kufer* (nonbelievers).

Religion is something people generally take so seriously that they have fought and died in the name of a particular belief system since the dawn of mankind. Religious disputes are more personal and result in either hurt feelings or, in many cases, ruined friendships and even broken relationships within family. The potential for escalation is always present, and the propensity for violence among religious followers has been documented for millennia. Ever since the organized religions came into effect, first religions there have had fights between followers. Some of these struggles have spread into a global scene and have taken the lives of millions.

In World War II, Hitler used Christianity as a tool to get the German people on side and engage with mass torture and murder. Hitler did not commit these acts in the name of Christianity, but he took a political move to convince the people that he was committing such act by wish of God. Nazi soldiers were committing these acts in the name of the religion. Regardless of Hitler's motives, the Holocaust happened because of Christianity. Hitler was a Christian, and with the help of the Catholic Church, he did his work in the name of God. Take the Thirty Years' War; religion was a motivation for the war as Protestant and Catholic states fought over what is now modern-day Germany and Italy. Involving most of the European countries, it was one of the most destructive conflicts in European history with an estimated 3 to 11.5 million casualties. The term *jihad* in Islam translates as "struggle" or "resisting" and is meant to refer to the defense of the religion against oppressors. Jihadists have been active in armed military actions for twelve centuries and have claimed millions of lives in the name of Islam.[33]

It is worth noting that a group of world religious leaders from the Buddhist, Protestant, Catholic, Orthodox Christian, Jewish, Muslim, and many other faiths met in Geneva, Switzerland, during October 1999. They issued a document, the Geneva Spiritual Appeal, asking political and religious leaders and organizations to ensure that religions are not used to justify violence in the future. Delegates believed that many of the conflicts have religious elements.[34] What is the outcome afterward? Nothing. Religious conflict, violence, and war have not stopped until today.

We may conceptualize three possible concepts of war that a religion promotes: (a) all violence and killing is wrong (the pacifist view); (b) wars are fought according to rules because they are perceived to be in the interests of justice, fight against injustice (a just war view); and (c) God asks and commands its followers to make wars on those who do not believe in that religion and who pose a threat to those who do believe (a holy war view).[35]

Nonviolence War

Primarily, there are three world religions that support nonviolence (ahimsa) wars. They are Hinduism, Buddhism, and Sikhism. The principal idea behind nonviolence is that the good in nonviolence is permanent, as opposed to violence, where good is temporary and the evil is permanent. These three religions explain the idea as follows:

Hinduism. In this religion's writings, ahimsa has been considered the highest duty. Jainism, which grew out of Hinduism, emphasizes that people should strive to become detached from the distractions of worldly existence and that the practice of ahimsa is an essential step on the way to personal salvation. The followers of Jainism even believe that killing of lives is a sin. The Hindu scripture, the Bhagavad Gita, tells a different story of Arjuna, who learns that it's his duty to fight as a member of the soldier caste. Arjuna is told by his chariot driver, Krishna, who is really the god Vishnu in human form, "Even without you, all the soldiers standing armed for battle will not stay alive. Their death is foreordained" (Bhagavad Gita 11:32–3).

In the story, Arjuna overcomes his doubts and fights, even though he knows it means killing some of his own family. Strict rules, however, are laid down for war, such as cavalry only goes into action against cavalry, infantry against infantry, and so on. The wounded, runaways, and all civilians are to be respected. This is the idea of a war, just for a war.

Buddhism. Buddhism was developed and spread from the teaching of Siddhartha Gautama, called the Buddha, who believed that human suffering could be overcome by following a particular way of life. The principal tenet of Buddhism is "Hatred is never appeased by hatred in this world; it is appeased by love" (Dhammapada 15).

The first precept of Buddhism is "nonharming" (ahimsa): Buddhism rejects violence. Many Buddhists promote the idea that it is better to be killed than to kill. Some Buddhists have been very active in promoting peace, particularly during the Vietnam War (1961–1975). For example, many Buddhist monks burned themselves to death in self-sacrificing protest against the Vietnam War. Buddhism perhaps has the best record of all religions for nonviolence. However, there are exceptions too. Buddhists in Sri Lanka have been criticized for oppressing the Tamil minority, who mostly belong to Hindu religion, whose origins are in Southern India.

It is worth noting that Confucianism and Taoism, which both developed in China, also share similar principles with Buddhism. For example, they seek to adjust human life to the inner harmony of nature (Confucianism) and emphasize mediation and nonviolence as means to the higher life (Taoism). The founders of these religions, Confucius and Lao Tzu, lived in the same period as Buddha, the sixth century BC.

Sikhism. The most important thing in Sikhism is the internal religious state of the individual. Sikhism stresses the importance of doing good actions rather than merely carrying out rituals. The Sikhs are not essentially violent but militant, where militancy means not only violence in actions and reactions alone but also an aggressive and passionate stand for the cause of their religion and the gurus. Guru Nanak, the first Sikh guru, wrote this hymn: "No one is my enemy. No one is a foreigner with all I am at peace. God within us renders us incapable of hate and prejudice."

Guru Nanak also emphasized the importance of nonviolence and the equality of all humans irrespective of their religious doctrine. But this pacifist theme has a different interpretation when Sikhs face persecution. The sixth guru said, "In the guru's house, religion and worldly enjoyment should be combined—the cooking pot to feed the poor and needy and the sword to hit oppressors." The tenth and last guru, Guru Gobind Singh, was a general as well as a guru. To strengthen the courage and military discipline of the Sikhs at a time of great persecution, he organized the Khalsa—the Sikh brotherhood.

Guru Gobind Singh expressed the idea of just war as follows: "When all efforts to restore peace prove useless and no words avail. Lawful is the flash of steel, it is right to draw the sword."

In all counts, the central teaching of Sikhism is respect for people of all faiths, not the violence against other religions.

Wars and Holy Wars

Judaism. Peace is the central teaching of rabbinic Judaism, but it is not a pacifist religion altogether. The idea of holy war is recorded in the Hebrew Bible, but in different ways, such as how the holy war was not about making others Jewish; rather, it was about their survival. For example, it is found in the Old Testament that said, "They shall beat their swords into ploughshares and their spears into pruning hooks: nation shall not lift up sword against nation, neither shall they learn war any more" (Old Testament, Isa. 2:4).

It is to be noted that revenge and unprovoked aggression are condemned, but self-defense is justified. Jews have been victims of dreadful persecution at the beginning of Islam and later at the hands of Christians for nearly two thousand years, which resulted in the Holocaust during the Second World War. On the other hand, defending modern Israel and dealing justly with the Palestinians place Jewish people in difficult dilemmas, violence, aggression, and peace.

Christianity. During the two-thousand-year history of Christianity, all kinds of wars have been witnessed. Jesus's teachings in the Sermon on the Mount (the New Testament, Matt. 5–7) are very clearly nonviolent. For example, "Blessed are the peacemakers, for they shall be called the children of God" (Matt. 5:9) and "Love your enemies" (Matt. 5:44).

Christian church until the Roman emperor Constantine (274–337) made Christianity the official religion of the empire. This is the teaching of pacifism, but later, it took a turn toward the development of the kind of war that resounded like the Just War. This suggests that politics and religion came together to endorse the idea of war. During the Middle Ages, the Crusades were fought mainly to recover the Holy Land (the area between the Mediterranean and the River Jordan) from Muslim rule. This was indeed the terrible cruelty that happened during the Crusades. It is also a fact that Christians persecuted heretic people (people who did not accept the official teachings of the Christian church) and non-Christians (such as Jews).

The majority of Christians supports the idea that war is regrettable but unavoidable and should be fought as a "Just War" rule. Pacifists are a minority group of Christians in the larger denominations (Roman Catholic, Church of England, Methodist, etc.). The Quakers, Mennonites, Amish, and Hutterites together make up the historical "peace churches" with a long tradition of pacifist belief and action.

Islam. The principal tenet of Islam is "submission" or "surrender" to the will of God (Allah). The Quran teaches that God is in control of everything that happens and that the Muslims do the will of Allah rather than follow an individual path through life. It is intriguing to note that Islamic teaching is often misunderstood, particularly on the matter of jihad. The question is, what does jihad mean? Islamic scholars say that jihad means to "strive" or "struggle" in the way of God.

Muslims use the word *jihad* to describe three different kinds of struggle: (1) a believer's internal struggle to live out the Muslim faith; (2) the struggle to build a good Muslim society; and (3) holy war—the struggle to defend Islam, with force if necessary.

All faithful Muslims are thus involved in a jihad, which is largely nonviolent. But there is another jihad (a war) that is commanded by Allah. The Quran says, "There shall be no compulsion in religion." On the contrary, it is also found in the Quran: "I will cast terror into the hearts of those who disbelieve. Therefore, strike of their heads and strike off every finger of them" (Quran: 8.12). "Oh you who believe! Murder those of the disbelievers and let them find harshness" (Quran 9:123). "Slay the idolaters wherever you find them" (Quran 9:5). "Fight those who do not believe in Allah and the last day . . . and fight People of the book, who do not accept the religion of truth (Islam) until they pay tribute by hand, being inferior" (Quran 9:29). These are the examples of Quranic verses in favor of war and jihad in Islam.[36]

Religious Oppression and Conflict

Killing in Religion

> *More wars have been waged, more people killed, and more evil perpetrated in the name of religion than by any other institutional force in human history. The sad truth continues in our present day.*
>
> —Charles Kimball

Violence and killing by the name of religion has been a curse upon humanity for centuries. It is also a fact that religious violence is tearing religions apart internally and causes harm to others.

History of modern religions suggests that Islam waged religious wars, forcing conversion to Islam as a means to spread its Islamic faith. In retaliation, Christians waged wars known as Crusades, spanned between the eleventh and thirteenth centuries. Moreover, the religion wars in Europe between different Christian sects took place in the sixteenth and seventeenth centuries. Six million Jews were exterminated because of anti-Semitic ideology. In the Arab-Israeli conflict, thousands have been killed. Many more will die if the long-lasting religious conflict between Muslims and Jews is not resolved by negotiations. These are only a few examples of violence, atrocities, and killings in the name of religion.

Christianity was once the world's most violent religion. Much of Christianity's killings were in the past. What Christians practiced centuries ago, Muslims are practicing now. Jihad and Sharia are parts of the core doctrines of Islam. Interestingly, killing for Islam is not a modern idea, and it will never end until some sort of reformation takes place within the religion. The truth is, medieval Christianity was even more violent, but it has since reformed, and now modern Christianity becomes more tolerant and peaceful than medieval times, which is a good move.

There has always been debate on the issue about who has been the worst killers and murderers throughout history—atheist regimes or religious regimes. Religious people claim that the most killings in history have been done by atheist regimes. There are examples that suggest that many killings are termed as nonreligious killings, but they are not. A group of world religious leaders from the Buddhist, Protestant, Catholic and Orthodox Christian, Jewish, Muslim, and many other faiths met in Geneva, Switzerland, during 1999. Delegates believed that many of the fifty-six conflicts have religious elements. Thus, they issued a document, "The Geneva Spiritual Appeal," asking political and religious leaders and organizations to ensure that religions are not used to justify violence in the future.[37]

The Israel-Palestine conflict became an ongoing matter in the Middle East between Jews and Muslims. How many people have been killed in the Israel-Palestine conflict? How many more will die, and how long will this conflict continue? Why can't either side come to a resolution? While some of the questions can be answered with numbers and facts, others are very complex and lack any real answers. Whatever it is, the situation in that region is extremely grim. As Jimmy Carter (2007) stated, the instability in the Middle East is a persistent threat to global peace. It is the incubator of much of the terrorism that is of great concern to Americans and other nations.

Israel killed more Palestinian civilians in 2014 than in any other year since the occupation of the West Bank and Gaza Strip began in 1967, a UN report has said. Israel's activities in the Gaza Strip, West Bank, and East Jerusalem resulted in the deaths of 2,314 Palestinians and 17,125 injuries, compared with 39 deaths and 3,964 injuries in 2013.[38] Currently, Israel controls 98 percent of Palestine and continues to expand its Jewish settlements in the West Bank and East Jerusalem. Meanwhile, according to the United Nations, there are now more than five million Palestinian refugees living in camps who hope to return someday to their land. The reality suggests that a largely disproportionate number of Palestinians have been killed or forcibly displaced as well as have lost huge lands they once owned. Moreover, continuing the trend of disproportionate death ratios, on average, twenty-eight Palestinians have been killed in 2014 for each

Israeli death. These statistics are drawn from UN sources and the Israeli human rights organization B'Tselem.[39]

What is the lesson learned from all these historical and recent happenings? Religions should act in a way that those who take religion for personal guidance are free to do so, but they must also abide by the laws that separate state and church, mosque, synagogue, temple, etc. Religious dogma proclamations should not be forced on anyone. Under these conditions, religious freedom may go hand in hand with personal freedom. This is a practical approach and perhaps the key theme in promoting all religions' fundamental tenets of brotherhood, sisterhood, compassion, harmony, and peace, which, in turn, may bring an end to the killing of people in the name of religion.

Honor Killing

According to Human Rights Watch, "Honor killings are acts of vengeance, usually death, committed by male family members against female family members, who are held to have brought dishonor upon the family. A woman can be targeted by her family for a variety of reasons, including: refusing to enter into an arranged marriage, being the victim of a sexual assault, seeking a divorce, even from an abusive husband or committing adultery. The mere perception that a woman has behaved in a way that 'dishonors' her family is sufficient to trigger an attack on her life."[40]

Honor killings had been in existence since ancient Roman times. It was the senior male within a household who had the right to kill an unmarried but sexually active daughter or an adulterous wife. Honor-based crimes were also practiced in medieval Europe by the Jewish community where Jewish law allowed killing by stoning for an adulterous wife and her partner. However, honor killings are not associated with particular religions or religious practice; rather, they have been across religious people such as Christian, Jewish, Sikh, Hindu, and Muslim communities. At present, the practice is most commonly linked to regions in North Africa, the Middle East, and South Asian countries.[41] Out of the five thousand honor killings that occur internationally each year, about one thousand happen in India, and one thousand take place in Pakistan, according to international digital resource center Honour-Based Violence Awareness (HBVA).[42]

Lindsey and Sarah (2010) looked at honor crimes in the way that is understood as a product of a culture that stresses honor and purity of women inside the family. Honor is a concept that defines one's self-worth through one's behaviors and actions and that those behaviors are in line with modesty and cultural traditions of the community. In Islamic communities, family honor is directly linked to the purity and chastity of the daughters, wife, sisters, and in-laws within the family unit (Devers and Bacon 2010). It is observed that when there is a suspicion or actual evidence is found that a female member of the household has dishonored the family and clan, harsh and cruel methods are used such as stoning, lashing, and forced marriages to amend for the dishonorable act. It's very shocking to note that murdering the perpetrator (i.e., the female) can be perceived as a reputable way to salvage family status and reputation. Subsequently, men in the direct family lineage, such as a father, brother, grandfather, or uncle, commit the majority of honor crimes; victims are most commonly women.

Mothers also commit honor crimes, especially when they think that it is essential to protect the family honor because of their daughter's premarital sexual activity or pregnancy (Human Rights Watch 2006). It is also a fact that in some cases, the female family members act indirectly in honor killing because they are encouraged by the head of the household. Most convincingly, women are the target for honor killing because of the fact that they have voluntarily violated the moral code or because they are the victims of sexual violence. Honor killings are very common in Islamic countries where honor and purity of women are stressed too strongly. Beheading the women, burning women to death, stoning women to death, stabbing women, electrocuting women, strangling women, and burying women alive for the "honor" of their families are extremely barbaric and very shameful and, of course, inhuman. It is one of the last great taboos that the murder of at least twenty-thousand women a year, worldwide, is in the name of honor. While it is almost impossible to provide detailed statistics of honor killings, we can cite some of the vivid, horrific examples that are shocking to the highest degree by human perception. (See annexure 4: "Vivid Horrific Cases of Honor Killings.")

Robert Spencer of Jihad Watch commented that honor violence or honor killing is not just a tribal custom; rather, it is Islamic law. He pointed out that a manual of Islamic law is certified as a reliable guide to Sunni orthodoxy by the most respected Al-Azhar University. The guide says, "Retaliation is obligatory against anyone who kills a human being purely intentionally and without right." However, "not subject to retaliation" is "a father or mother [or their fathers or mothers] for killing their offspring, or offspring's offspring" (*'Umdat al-Salik* o1.1–2). What it means is that the law is that a Muslim father or mother who kills his child for "honor" faces no legal penalty under Islamic law.[43]

Phyllis Chesler and Nathan Bloom wrote an article in the *Middle East Quarterly*, where the authors explained the following: Although Islam does not specifically endorse killing female family members, some honor killings involve allegations of adultery or apostasy, which are punishable by death under Sharia (Islamic law). Thus, the belief that women who stray from the path can be rightly murdered is consistent with such Islamic teachings. The refusal of most Islamic authorities to unambiguously denounce the practice (as opposed to merely denying that Islam sanctions it) only encourages would-be honor killers.[44]

In an article, Gautaman Bhaskaran has attempted to explain the perception of honor killing in India. Like many others, he mentioned that the concept of killing to restore honor of the family or community or tribe is more prevalent among Muslims but that the crime also occurs in Hindu and other communities in India. In some states of India, men and women are hanged in full public view in response to disobeying parental wishes—marrying outside their caste in particular. India is still a traditional society, and women's rights are deeply ingrained in mostly patriarchal communities where men with landed properties and other economic resources become superior to women, especially in rural areas.[45] According to the United Nations, one in five cases of honor killings internationally every year takes place in India. This is a staggering number. The *Times of India* reported that in India, honor killings are more prevalent in Punjab, Haryana, and western Uttar Pradesh. These honor killings occur not only within the Muslims but also among Sikhs and Hindus.[46]

A study conducted by research corporation Westat and commissioned by the US Department of Justice identified four types of honor violence: forced marriage, honor-based domestic violence, honor killing, and female genital mutilation. The report estimated that twenty-three to twenty-seven honor killings per year occur in the United States. However, the cases are often unreported because of the shame they can cause to the victim and the victim's family. In most cases, such incidences are kept hidden under cases labeled as domestic violence motivated by radical interpretation of Islam, according to Fox News. The report also noted that 91 percent of victims in North America are murdered for being too Westernized, such as sporting sleeveless or wearing jeans, befriending boys, and choosing to be free from family traditions. And the victims are daughters, eighteen years or younger. Ironically, for every honor killing, there are many more instances of physical and emotional abuse, all in the name of religion. Honor killing and violent crime are even more prevalent in other parts of the Western world. Some report suggests that there are eleven thousand cases of honor violence recorded in the United Kingdom in the last five years. Cases of honor killings and violence have also been documented in Western countries, including Canada.[47]

There have been a number of high-profile honor killings in Canada; the most notable are the murder of Kaur Sidhu, the murder of Kanwaljeet Kaur, the murder of Amandeep Atwal, the double murder of Khatera Sadiqi and her fiancé, the murder of Aqsa Parvez, the murder of Aysar Abbas, the murder of Shaher Bano Shahdady, and the Shafia family murders, among many others. As such, honor killings have become a pressing issue in Canada, and the Canadian citizenship study guide describes it by saying, "Canada's openness and generosity do not extend to barbaric cultural practices that tolerate spousal abuse, 'honor killings,' female genital mutilation, forced marriage, or other gender-based violence."[48]

"Once social change begins, it cannot be reversed. You cannot uneducate the person who has learned to read. You cannot humiliate the person who feels pride. You cannot oppress the people who are not afraid anymore," said Cesar Chavez. This is a powerful message indeed.

Someone has said, "It is a tragedy, a horror, a crime against humanity. The details of the murders—of the women beheaded, burned to death, stoned to death, stabbed, electrocuted, strangled, and buried alive for the 'honor' of their families—are as barbaric as they are shameful."[49]

Women of the world have this to say: *We have a right to a secured life free from the horrors of honor killings, rape, and mutilations!*

Honor killings are heinous crimes against humanity. Let humanity speak for itself. These crimes go against all humanity. What kind of religion allows such barbaric and horrified acts yet claims religion of peace? According to Human Rights Watch:

Although the entire spectrum of criminality has been woven throughout the annals of the History of this world wherein crimes against women and young girls have accumulated to hundreds of thousands of wrongful deaths and multiple acts of horrific abuses the time to put a final end to "Crimes against Humanity" now takes center stage. Regardless of cultural traditions, religious beliefs or any other practice which have consistently condoned horrific violations of Human Life in any form whatsoever and/or for whatever reason such crimes have been allowed to be perpetrated "Crimes against Humanity" must now be legitimately Halted! It's everyone's responsibility to report crimes of "Honour Killings," Rapes and Mutilations.[50]

Child Marriage

United Nations Population Fund categorically said that child marriage is a human rights violation. Although there are laws against child marriage, the practice remains widespread, mainly because of persistent poverty and gender inequality in the developing countries. One in every three girls is married before reaching age eighteen, and one in nine is married under age fifteen in many countries across the globe.[51] Each year, fifteen million girls are married before the age of eighteen. That means twenty-eight girls every minute. More than seven hundred million women alive today were married before their eighteenth birthday. That is the equivalent of 10 percent of the world's population. If the current rate of child marriage continues, an additional 1.2 billion girls will be married before eighteen years of age by 2050. Child marriage ends up with disempowerment of girls, making them dependent on their husbands and depriving them of their fundamental rights to health, education, and safety. Nonetheless, these underage girls are ready neither physically nor emotionally to become wives and mothers when married at childhood and are subject to experiencing greater risks of dangerous complications in pregnancy and childbirth, such as being infected with HIV/AIDS or suffering from domestic violence.[52]

Most of the major religions have some influence on the child marriage. In Christianity, the ecclesiastical law forbade marriage of a girl before the age of puberty. Prior to 1917 Code of Canon Law, the minimum age for a valid marriage was puberty, or nominally fourteen for males and twelve for females. However, the 1917 Code of Canon Law raised the minimum age for a valid marriage at sixteen for males and fourteen for females, which remained the same in the 1983 Code of Canon Law. Jewish scholars and rabbis have strongly discouraged girls' marriages before the onset of puberty. But they also recommended that the girls between age three and twelve years (which is the legal age of consent, according to Halacha) can be given in marriage by her father. Likewise, Islamic marriages have permitted marriage of girls at the age of ten or below, based on Sharia Law, which is based in part on the life of and practices of Prophet Muhammad, described in Sahih Bukhari. The third wife of Prophet Muhammad, Aisha, was only six years old when she was married, and she was nine years when the husband consummated the marriage.

Hindu Vedic scriptures mandated the age of a girl's marriage to be adulthood, which is defined as three years after the onset of puberty that is sixteen to eighteen years of age. In the earliest recorded history of India, young women and men practiced a liberal life concerning love, romance, and marriage, including choosing a partner without any fear of scandal. So child marriage (Bal-Vivaha) was not widespread in the Indian culture. However, from the Middle Age, the political system brought changes and modified the Indian society to a great extent. There had been significant changes in the lifestyle and opinion of the people in India. As a result, women lost their rights and had to obey rules and respect the code of behavior, including the system of marriage. They were compelled to obey laws regarding family discipline and the honor of their clan.[53]

At present, the legal age for marriage in India is eighteen years for women and twenty-one years for men. Any marriage of a person younger than this is banned under the Child Marriage Prevention Act of 1929. Although illegal, the practice of child marriage is widespread and accepted by the majority of Indian society, especially in the rural areas of the country where girls are seen as a liability and burden for the family. The reason is that it is the girl's parents who are obligated to pay the dowry at the event of the daughter's wedding. Thus, the longer the parents wait to get the daughter married off, the more they will have to spend on the daughter, including the amount of dowry to be paid to the future in-laws. However, many argue that the maturity of any religion should be judged based on its dynamism that is open to change from time to time in response to reality. If a religion sees and admits something wrong in its values, principles, and practices, then can it stop and still be religious? It's a good thinking indeed.

Hinduism has changed with the changing time in the matters of sati, child marriage, widows burned alive, remarriage of a widowed woman, and lenient attitudes toward untouchables, etc. Who knows? The caste system will be eliminated someday, may not be in the near but in the far future. Let us hope that all major religions in the world would be rational and dynamic in bringing the change with the changing time and make necessary reforms in the values, principles, and practices in their respective religions. This, in turn, will save millions of girls' lives from being tortured (physical and psychological) and forced child marriage in the name of religion and cultural taboos.

Approximately 37 percent of girls in the rural areas and 21 percent girls in urban areas of Pakistan are married as child brides. It is also estimated that 30 percent of girls nationwide are married before the age of eighteen years.[54] According to reports from the Ministry of Women's Affairs and NGOs, 57 percent of women in the twenty-five-to-forty-nine age group are married by the age under sixteen, some of them as young as nine.[55]

Child marriage is a serious human rights violation issue that affects children's and women's rights to health, education, equality, nondiscrimination, and to live free from violence and exploitation. These are rights enshrined in the Universal Declaration of Human Rights, the Convention on the Rights of the Child (CRC), the Convention on the Elimination of All Forms of Discrimination against Women (CEDAW), as well as other international and regional human rights organizations. Michelle Bachelet, MD, executive director of UN Women, said, "No girl should be robbed of her childhood, her education and health, and her aspirations. Yet today, millions of girls are denied their rights each year when they are married as child brides."[56]

Child marriage may also be considered slavery in some cases, for example, when children are bought and sold in the name of marriage for the purposes of sexual exploitation or when they are traded or trafficked into forced marriages. These are the cases that fit internationally recognized definitions of slavery or similar practices. By any measure, it is a crime against humanity at large.

The Humanist View

In recent times, religion has played not only a controversial but also a negative role in many societies, particularly in the West. Many people have consciously rejected the notion of a spiritual and sacred religion or god. However, this trend does not necessarily mean the rejection of ethical principles of mankind. Some people have developed a philosophy of humanism, which is based on humanitarian ideals, such as individual responsibility for one's actions, respect for others, helping and cooperating for the common good, and sharing available resources. In fact, this is a right and meaningful approach to global human advancement and overall progress toward human bondage for peaceful and harmonious coexistence.

Humanists accept the Golden Rule, a term used by Confucius, which says, "Do as you would be done by" or "Treat others as you would wish them to treat you." We all should see the natural, logical, and rational conclusion of such a principle to reject all war, oppression, and violence. Humanism is the only road to reach mankind's ultimate goal—love and peace with the rejection of war and violence. Who knows? Without religious faith in modern times, there would have been no 9/11 attacks, no Israeli-Palestinian conflict, no Troubles in Northern Ireland, no India-Pakistan war, no Taliban uprising, no Iraq-Iran War, no Syrian war, no creation of ISIS, and no disputes over the texts in holy scriptures.

Problems with Religion

Since the beginning of the new millennium, religious understanding and the concepts of God have changed very little. Both religions and the ideas of God have been developed within cultural contexts with the interpretations based on race, knowledge, environment, and economy. Religion is motivated by fear and insecurity and thus feels so alone and helpless that they need something or someone to keep faith for salvation. Life is a cruel, sadistic torment in many ways in many countries in the world, which, in turn, raises questions whether or not there is any good god in charge of everything and whether or not religion has any value to their lives. The proponents and opponents of various faiths have been engaged in the religious tug-of-war and, as a result, have destroyed cultures and invaded individual rights. The differences of beliefs have led to a horrendous amount of suffering of the people. The people of faiths different from the ruling religion have been persecuted mercilessly. Thankfully, some parts of the world have attained a certain level where persecuting and prosecuting of the members of minority religions and nonreligious freethinkers for blasphemy and heresy are not the general norm. However, in many countries, freethinkers, secularists, agnostics, and atheists remain outcasts, even though many of the world's greatest thinkers and distinguished scholars belong to those groups.

Among the many atrocities committed in the name of God and religion over the centuries experienced the practice of human sacrifice. It was, and is still today, claimed that superiority of a religion is determined by miracles and the number of people who are willing to die for the sake of religion.

Every religion has had its martyrs. For example, in Hinduism, limbs wither after years of painful persistence in vows to his deity, and believers who cast themselves under the wheels of the car of Jagannath establish the soundness of their creed; in Christianity, Jews (who, for centuries, bore the fiercest insults of the world) were persecuted, hunted, and killed by every conceivable torture for their denial of the incarnation, resurrection, and ascension and in their rejection of Jesus Christ. Muslims have regularly martyred themselves—if one dies in a holy war, he is destined to go to heaven. Will a Christian or a Jewish or a Hindu or a Buddhist agree that Islam is the true faith? The sufferings of the apostles do not prove anything beyond their own belief. Martyrdom gives nothing except the passionate feeling of the believer.[57]

Many people blame organized religions for mankind's problems. Others believe religion has failed to answer the vital questions regarding the purpose for human life. Modern civilization has achieved awesome advancements in technology, science, and knowledge. Ironically, humanity, within religion, is experiencing a rapid moral decline and is ignorant of the answers to the multifaceted problems we have been facing today. With thousands of religious figures proclaiming that their brand of religion should be followed, which message is correct? So there is confusion. The result is not surprising. Millions of believers are on the verge of giving up on religion altogether. Some go at it alone as independents, thinking that traditional religion has failed them. Many believers profess that religion has lost credibility.

According to Pew Research, societies have become increasingly less attached to religion. It does not mean that we don't believe in God; rather, we prefer to label ourselves as unaffiliated or perhaps spiritual but not religious. Religion is an institution that actually hinders spirituality.

There is no problem in having faith in God, Allah, or Bhagaban or whoever we name them. But the problem arises when we put human-made rules and interpretations on these faiths. Organized religions are the most unfavorable things to mankind. In fact, all organized religions provide a tool for those in power to manipulate millions to act against their own self-interests and block freethinking. People in power use religion to bring people to act like absolute fools. People lead their lives under the control of religious rules that were written by some people in the holy books, which cannot provide factual evidence to support them. But the irony is that as long as billions of people continue to follow these human-made religious rules, humanity is subject to be jeopardized.[58]

Religion promotes tribalism. By this, it is meant that religion divides insiders from outsiders. Religious followers are taught to treat outsiders with suspicion, such as, "Be ye not unequally yoked with unbelievers," says the Christian Bible. "They wish that you disbelieve as they disbelieve, and then you would be equal; therefore take not to yourselves friends of them," says the Koran (Sura 4:91). These kinds of religious teachings obviously discourage or even forbid the kinds of friendship and intermarriage that help clans and tribes become part of a larger whole. Furthermore, outsiders of a particular religion are seen as enemies of God or the agents of Satan who lack morality and are not to be trusted. One can imagine what damage it can make to the people and societies.[59]

Religion takes back believers to the Iron Age. The Iron Age was a time of rampant superstition, ignorance, inequality, racism, misogyny, hate, violence, and slavery. Women were not only treated badly but also considered inferior to men who are in possession of women. Warlords were constantly engaged in warfare. People sacrificed living animals, agricultural products, and enemy soldiers as burned offerings intended to appease gods. Holy books like the Bible, Torah, Koran, and others preserve and protect the old cultures and practices by the name of god or goddesses. Religious believers find excuses for demonstrating the sense of superiority, bigotry, wars, and many other destructive pursuits—that all these are written text revealed by God. While moral consciousness by virtue of humanity is emerging—that is, the Golden Rule—religious believers are reluctant to move away from orthodox belief; rather, they are anchored to the Iron Age. There's also the fact that religion makes a virtue out of faith, which is the final word. When science and technology are taking over the territory once held by religion, ancient, traditional, orthodox religious beliefs require strong defense against open and modern information, which is a threat to the belief. To stay on the course, religion provides commendable teaching and vigorous training to the believers to practice self-deception, not to accept contradictory information, and to just trust religious authorities and, more importantly, not to think but only to believe.[60]

Religion teaches helplessness. Let us see how it is. God has given the problem, and God will solve the problem. These statements are commonly heard quite often. A point to be noted is that we hardly recognize the intrinsic relationship between religiosity and resignation. Poverty, accidents, disease, and life's problems are given by God as believers think, rather than think that those are the results of bad decisions of people or bad systems. This attitude obviously is harmful to the society at large, as well as individuals. Since the major religions came into existence, common people had little power to change social structures through either scientific or technological innovations.[61] Any critics of religion create a lot of problems. For example, when we say religion is outdated, harmful to the individual, harmful to society, an impediment to the progress of science, a source of immoral acts or customs, and a political tool for social control, we are subject to punishment or even death threats. Religious problems also include compulsion, delusion, honor killings and stoning, blood sacrifice, genital mutilation, holy war, religious terrorism, suppression of scientific results and progress, racism, corruption, and so on.

The single most dangerous problem with religion is the fact that it demands that we give up our most important human asset, the ability to question. It demands that we simply believe but not question any of the texts of the scripture. Freethinking is a virtue, a birthright. But religion takes away this freedom from us. Insofar as social development is concerned, nothing is of greater importance than the human function of questioning. "Questioning led to the development of civilization," said Vladimir Pozner. I think it is synonymous to what Benjamin Disraeli said, "Where religion begins, knowledge ends."

Let me remind the readers of some of the thoughts of world-famous, distinguished people about religion, although I don't assume that many readers are unaware of them. These thoughts are very much related to the problem with religion.

> *When two men of science disagree, they do not invoke the secular arm; they wait for further evidence to decide the issue, because, as men of science, they know that neither is infallible. But when two theologians differ, since there are no criteria to which either can appeal, there is nothing for it but mutual hatred and an open or covert appeal to force.*
> —Bertrand Russell

> *Religion is regarded by the common people as true, by the wise as false, and by the rulers as useful.*
> —Seneca the Younger

45

I think there's something very evil about faith . . . it justifies essentially anything. If you're taught in your holy book or by your priest those blasphemers should die or apostates should die (anybody who once believed in the religion and no longer does needs to be killed) that clearly is evil. And people don't have to justify it because it's their faith.

—Richard Dawkins

My feeling about religion is at par with Albert Einstein, who said he believes in Spinoza's god. The nature is mysterious to us because of its very creation and attributes—beauty and ugliness, good and bad, love and cruelty, creation and destruction, birth and death, and so on. These are the sources of human knowledge. Our insight about the mystery of nature and life per se, coupled with fear, insecurity, pain, and survival, has given rise to religion. Know that the unknown manifests itself as the highest wisdom that our faculties can comprehend. This feeling is the center of true religiousness. My understanding is that we don't have to follow a traditional religion or to practice its rituals to become a good, moral, disciplined, and decent human being.

Will Religion Disappear?

Atheism is on the rise around the world, so does that mean spirituality will soon be a thing of the past? "The answer to this is far from simple," said Rachel Nuwer. More than a billion people worldwide do not associate with any religion, and millions are atheists who believe that life ends at death and that there is no god, no afterlife, and no divine plan. It's a rational viewpoint that could be gaining momentum. Statistics suggest that there are more atheists across the globe today than ever before, both in absolute number and in proportion, said Phil Zuckerman, a professor of sociology and secular studies at Pitzer College in Claremont, California, and author of *Living the Secular Life*.[62] A Gallup International Survey[63] of more than fifty thousand people in fifty-seven countries reveals that the number of individuals claiming to be religious fell from 77 percent to 68 percent between 2005 and 2011, while those who self-identified themselves as atheists rose by 3 percent, bringing the world's estimated proportion of adamant nonbelievers to 13 percent.

While atheists certainly are not the majority, could it be that these statistics show that the numbers will increase in the future? Does it tell that religion will disappear someday, assuming that this global trend continues?

It's not possible to predict the future though, but we can examine and make our best guess on what we know about religion given the inquisitiveness about why it evolved in the first place and why some people chose to believe in it and others don't—may predict as to how our relationship with the divine might play out in decades or centuries to come. Scholars are still trying to find out the complex factors that drive an individual or a nation toward atheism and agnosticism. A part of religion's appeal is that it offers security in an uncertain world. So not surprisingly, countries that report the highest rates of population that disassociate with any religion (such as atheism and agnosticism) tend to be those that provide their citizens with relatively high economic, political, and existential stability. Zuckerman thinks that these securities are likely to diminish religious belief. He also thinks that capitalism, access to technology, and education seem to have correlation with religiosity in some populations.

Countries like Japan, the United Kingdom, Canada, South Korea, the Netherlands, Czech Republic, Estonia, Germany, France, and Uruguay (where the majority of citizens have European roots) where religion was important in the past now have the reduced belief rates in the world. These countries feature strong educational and social security systems and low inequality and are all relatively stable. Also, decline in belief seems to be occurring in many other countries, including places that are still strongly religious. Very few societies are more religious today than they were forty or fifty years ago.

"It's obvious that decline does not mean disappearance," said Ara Norenzayan, a social psychologist at the University of British Columbia in Vancouver, Canada, and the author of *Big Gods*.[64] Security for existence is very important than it seems. Insecurity, fear, suffering, and hardship in life instill religiosity. For example, in a moment, everything can change: a tornado can destroy a town; a doctor can issue a terminal diagnosis. When climate wreaks havoc and natural resources potentially grow scarce, then suffering and hardship could fuel religiosity. People want to escape from suffering, and when they find no ways to get out of it, they want to hold on to religious belief. "For some reason, religions seem to give meaning to suffering, much more so than any secular belief or reason that we know of," Norenzayan said.

Human beings naturally want to believe that they are a part of something bigger. Our minds search for the purpose and explanation of life. Norenzayan also said that although, with education, knowledge of science, and critical thinking, people might stop trusting their intuitions, the intuitions are there. Science is the knowledge with which many atheists and nonbelievers attempt to understand the natural world, but it is not an easy cognitive thing to conceive. Science is cognitively difficult, whereas religion is mostly something we don't have to learn because we already know what it is; just the belief. We experience that religious thought is the only path that has the least resistance. We need to change some fundamental things of our humanity to get rid of religion. This biological factor perhaps explains the fact that, although billions of people who are not affiliated with a church, mosque, or temple, many of them still believe in God and say they are spiritual.

It is true that the religious intellectuals, rationalist thinkers, and even the majority of the educated people who studied their respective religions with analytic minds could not, in the past and even today, denounce and reject religion altogether, as it was ingrained in their subconscious mind. This is the most powerful narcotic if it is injected to a person from childhood. This is exactly the brain chemistry that works for every religious individual. The uneducated and not fully rational common population believes and practices religious rituals without knowing what they believe in and practice. On the contrary, for the intellectuals and rational people of religious faith, it may be difficult to accept their holy book at the face value, and thus, they try to justify esoteric meanings in the verses and writings of the holy books. What choices do they have? They have to live within the circle of friends and relatives vis-à-vis within the community where they have inexorable pressure to be within the religion bondage; even many do not practice rituals.

In all human cultures, religions play an important role. They have evolved over millennia into countless variants as parts of the respective cultures, also functioning as an important factor of social coherence. Having easy explanations to existential questions, pacifying the fear of death, and keeping compact social binding through the provision of various sets of rules and practices, religions crucially contributed to the survival success of populations. Religions are often better than other meme complexes (such as science, for instance) at explaining the world's function on an emotional level. They provide answers to existential, emotionally appealing questions, creating an acceptable world model. The model provides a certain spiritual satisfaction regardless of its lack of consistency because of cognitive dissonance. A religion can spread regardless of the truth or falsity of its claims.

Some of the most powerful and elaborate meme complexes can be recognized as components of contemporary religious systems. Religions tend to consist of some basic core memes (e.g., the belief in a god or goddess, surrounded by symbiotic doctrinary memes, ethical systems, disciplined group behavior, etc.) and a wider cluster of related memes (religious narratives, interpretations, holy texts, symbols, etc.). This symbiosis has proved effective through time, as it is evident, even today, that religions still influence everyday life that affects the cohesion of large population groups. The "God meme" in most religions consists of a number of explicit commandments and pronouncements purportedly attributed to a god. Theistic memes, in general, are memes that regulate individual or collective behavior.[65]

People guess that religion will never go away, at least in the foreseeable future. Religion, whether it's built in and maintained through fear or love, reward or punishment, is very successful at perpetuating itself. Without this, religion would no longer be with the people. We need comfort in the face of pain and suffering and anguish that gnaw the heart. We think that there's some immortal, omnipresent, omnipotent, omniscient, and invisible being there above us to look after us when needed. When we face an ecological crisis, a global nuclear war, or an impending comet collision, the god would definitely emerge as savior, as believers understand. History of time is beyond limit. We exist for a short time, and in that time, we explore only a small part of the unlimited whole. We, the humans, are a curious species; we want to know, and we seek answers to many questions. So who knows what would happen thousands of years from now? Does religion ever exist among mankind or disappear forever, or is it just a temporary phenomenon in the history of mankind? Time will tell.

Chapter One Conclusion

I sing of equality. There's nothing greater than a human being nothing nobler! Caste, creed, and religion—there are no difference. Throughout all ages, all places, we're all a manifestation of our common humanity.

—Kazi Nazrul Islam

As we go all the way back to the dawn of human consciousness, we find religion everywhere in the world. Since primitive time, humans lived in a hostile world and encountered constant fear of nature's hostilities such as thunder, lightning, volcanoes, flood, etc., fear of disease, fear of animals, and so on. In search of securities, they created different gods to the fact that gods can give them comfort and courage to face calamities of life. Our ancestors conceptualized animism, which was the first bona fide religion whose roots go back to the Paleolithic period. Much later, during the Neolithic period, our ancestors began to form religions into an organized way, which emerged as a means to provide social and economic stability. For people in modern societies, religion is considered neither good nor bad but simply irrelevant, given the many alternative ways to find meaning in various forms of cultural pursuits, ethical and moral ideals, and lifestyles; all these lead to secularism and humanism at large.

Organized religion has been a prominent topic and has influenced and shaped all people's lives. There have always been believers and nonbelievers. Since the beginning, nonbelievers have been persecuted by inquisitions, prosecuted by witch trials, and murdered by stoning and crucifying for even questioning the "truth" about a supreme being. Skeptics in today's culture and many people think that organized religion does more harm than good. For example, some religions justify violence, including wars called crusades and jihad.[66]

Theories of religion are of many approaches such as philosophical, neurological, biological, psychological, social, scientific theories, and so on. However, in recent times, theories based on natural sciences, such as evolutionary biology and cognitive psychology, seem very popular. Literatures suggest that evolutionary theories of religion have two opposing schools of thought. According to Pascal Boyer (2001), religion is a by-product of the evolution of the human brain without adaptive functions, which is a natural phenomenon. The other theory is that religion is a product of biological and cultural evolution with immense adaptive value for group survival. This was suggested by Loyal Rue (2005). He thinks that religion is of utmost importance for the survival and well-being of the human species.[67]

There has been a long list of religious theories in modern time. However, for the scope of this book, I have presented only a handful of them in this chapter. Yet it would be good to provide annotated views of some scholars in the theories of religious studies, which is presented by Nathaniel Aminorishe Ukuekpeyetan-Agbikimi (2014) in the *Global Journal of Arts Humanities and Social Sciences* vol.2, no. 7. I would like to present very briefly the views of some of the scholars.

Karl Marx's Views of Theories of Religion

Karl Marx contemplated a strictly materialist worldview and saw economics, including class distinctions, as the determining factor of society. According to him, the dynamics of society was fueled by economics. He saw religion originating from alienation and aiding the persistence of alienation. He says, "Religion is supportive of a status quo in correspondence with his famous saying that religion is opium of the people."[68]

Edward Burnett Tylor's and James George Frazer's Views of Theories of Religion

The anthropologist Edward Burnett Tylor defined religion as belief in supernatural beings and stated that this belief originated as explanations to the world. Belief in supernatural beings grew out of attempts to explain life and death. Tylor's theory assumed that the psyches of all peoples of all times are more or less the same and that those explanations in cultures and religions tend to grow more sophisticated via monotheist religions like Judaism, Christianity, and Islam. James George Frazer followed Tylor's theories to a great extent, but he distinguished between magic and religion. According to him, magic is used to influence the natural world in primitive man's struggle for survival.[69]

Emile Durkheim's Views of Theories of Religion

Emile Durkheim saw the concept of the sacred as the defining characteristic of religion, not faith in the supernatural. He saw religion as a reflection of the concern for society, his interest in sociology as an academic discipline where he championed the importance of society—social structures, social relationships, and social institutions—in understanding human nature. He focused on the concept of the "sacred" and its relevance to the welfare of the entire community. From Durkhiem's point of view, religious beliefs are symbolic expressions of social realities, without which religious beliefs have no meaning. His approach gave rise to functionalist school in sociology and anthropology. Functionalism is a sociological paradigm that originally attempted to explain social institutions as collective means to fill individual biological needs, focusing on the ways in which social institutions fill social needs, especially social stability.[70]

Max Weber's Views of Theories of Religion

Max Weber developed theories from the sociological classifications of religious movements. He thought that the truth claims of religious movement were irrelevant for the scientific study of the movements. Weber acknowledged that religion had a strong social component but somewhat diverged from Durkheim. In *The Protestant Ethic and the Spirit of Capitalism*, religion can be a force of change in society. He said that modern capitalism spread quickly because of the Protestant worldly ascetic morale. Weber's main focus was not on developing a theory of religion but on the interaction between society and religion. Weber saw charisma as a volatile form or authority that depends on the acceptance of unique quality of a person.[71]

Sigmund Freud's Views of Theories of Religion Freud

Freud saw religion as an illusion. By illusion, he meant a belief that people want very much to be true. He attempted to explain why religion persists in spite of the lack of evidence for its tenets. Freud asserted that religion is a largely unconscious, neurotic response to repression. By repression, he meant that civilized society demands that we cannot fulfill all our desires immediately but that they have to be repressed. Rational arguments to a person who holds a religious conviction will not change the neurotic response of a person. He asserts that monotheist religions grew out of a homicide in a clan of a father by his sons. This incident was subconsciously remembered in human societies.[72]

Rudolf Otto's Views of Religious Theories

Being a psychologist and a theologian, Rudolf Otto (1869–1937) focused on religious experience, more specifically moments that he called numinous, which means "wholly other." He described it as *mysterium tremendum* (terrifying mystery) and *mysterium fascinans* (awe-inspiring, fascinating mystery). He saw religion emerge from these experiences. He further asserts that these experiences arise from a special nonrational faculty of the human mind, largely unrelated to other faculties, so religion cannot be reduced to culture or society. Some of his views, among others, that the experience of the numinous was caused by a transcendental reality are unverifiable and, hence, unscientific.[73]

Mircea Eliade's Views of Theories of Religion

Mircea Eliade's approach grew out of the phenomenology of religion. Like Otto, he saw religion as something special and autonomous that cannot be reduced to the social, economical, or psychological alone. Like Durkheim, he saw the sacred as central to religion, but differing from Durkheim, he views the sacred as often dealing with the supernatural, not with the clan or society. For him, the daily life of an ordinary person is connected to the sacred by the appearance called hierophant (a being who interprets and explains obscure and mysterious matters, especially sacred doctrines or mysteries).[74]

Clifford Geertz's Anthropological Theory

Geertz defined religion as (1) a system of symbols that acts to (2) establish powerful, pervasive, and long-lasting moods and motivations in men and women by (3) formulating conceptions of a general order of existence and (4) clothing these conceptions with such an aura of factuality that (5) the moods and motivations seem uniquely realistic. Geertz's definition, by far, has been the most influential anthropological definition of religion in the twentieth century.[75] However, religious anthropologists debate and reject the cross-cultural validity of these categories (often viewing them as examples of European primitivism). Anthropologists have considered various criteria for defining religion—such as a belief in the supernatural or the reliance on ritual—but few claim that these criteria are universally valid.[76]

In today's global world, we face unprecedented problems such as environment, globalization, ethnic cleansing, religious divisions, fundamentalism, radicalization, and so on. While many people argue that only science can lead to the solution of the problems, others reject this view. Rather, the universal unity between people and their belief systems is the main tenet to find rational solution to the problems mankind is facing, as majority of others think. The rational argument that establishing a strong unification of science, philosophy, and religion should be the centerpiece for discussion is aimed at finding a common ground.

In all religions, the most important theme is encrypted in humanity. All rituals and practices in religions are not so important than to find the inner power—the god in oneself—the true nature of mankind. This true nature is the sole purpose that would unify mankind of all religions and bring together all attributes called humanity. It may sound an impossible proposition at the moment, but the time will come when this proposition would be true. Only then will the unification of world religion be a reality, and this, in turn, would bring peace and harmony among mankind. The ultimate goal is to reconcile and unify all religions into one religion—humanity.[77]

According to divine command theory, things are right or wrong simply because the gods command or forbid them. There is no other reason. Socrates posed a famous question: "Are things right because the gods command them, or do they command them because they are right?" If things are right simply because the gods command them, then their commands are arbitrary—without reason. There are no good reasons for their commands. The gods then are like petty tyrants who just command things because they have the power. If we want to rationally justify morality, then we will have to do it in a moral theory independent of gods. We will have to engage in philosophical ethics and morals.[78] Instead, right and wrong are not to be understood in terms of God's will; morality is a matter of reason and conscience, not religious faith, and in any case, religious considerations do not provide definite solutions to most of the moral problems we face, said Rachelles. So morality can be defined as principles concerning the distinction between right and wrong or good and bad behavior.[79]

We must not forget that slavery is justified by both Christianity and Islam. The Torah, New Testament, and Quran are all full of situational ethics where it says that you can kill those different from you or who question your authority and that it's only a sin to kill others of your kind, Matt Dillahunty said in a lecture on the superiority of secular morality. His argument is that the secular sources of morality are based on reasoning that is "I will act a certain way to others because I wish them to act that way toward me and to an internal sense of right and wrong."[80] Someone said that the cultures of the world have their codes of conduct, even the ancient ones, and many of these codes predate the writing of the various scriptures and holy books such as the Bible, Torah, and Quran. We can learn things from the ancient and modern history, philosophy, ethics, and sociology from all over the world. We can see what worked well and what didn't. We have found that any group or culture has had all the answers to the challenges and problems we face in the changing world, and thus, we refuse to be tied to laws that should be abandoned and rules that do not work. Claiming to have all the answers in one religion or philosophy or ideology is a dangerous game, and claiming to be the one true path has led to some of the most horrific abuses and tragedies in human history, as someone has said. Albert Einstein said, "A man's ethical behavior should be based effectually on sympathy, education, and social ties and needs. No religious basis is necessary."[81]

The arguments I have considered above do not mean that Christianity, Islam, Judaism, or any other theological system is false; rather, they only show that morality remains to be an independent matter. The point is that there is morality without religion, meaning a morality that is not of dogmatic commands but of rational values and of unbounded respect for the life of the individual and of all people. With no religious affiliation, freethinkers do have freedom of conscience with self high-esteemed mental state to judge good and bad, which leads them to be moral people, because they consider reality through a variety of different perspectives without compelling them to believe false perceptions that are religiously defined.

CHAPTER TWO

Beyond Religion

No Religious Affiliation

People's lives without religion are not incomplete or unfulfilling. There are all attributes, characteristics, and values among nonreligious people. They are not immoral; rather, they have very strong ethical and moral principles, and even nonreligious people appear more tolerant, more law-abiding, less prejudiced, less vengeful, and less violent than their religious peers (Zuckerman 2014). Nonreligious people are very much capable to cope with life's troubles and adversities and can achieve highest fulfillment in personal and social arena. Religious nonaffiliation is nothing new. People in the past were suspicious of priests, imams, rabbis, and saints and questioned their teachings. For example, the Carvaka (a group of philosophers, materialist thinkers) rejected the supernaturalism of Hinduism. They were atheists and argued that there is no god or any afterlife. In China, the philosophy of Xunzi in the third century BCE promoted that there is no heaven other than the natural world and also that morality is not divine established but, rather, constructed by human. Jewish philosopher Kohelet of ancient Israel during the third century BCE suggested that all life is ultimately meaningless and that there is no life after death. In the ninth century, another Jewish philosopher, Hiwi al-Balkhi, blatantly questioned the divinity of the Torah, which caused many people to lose their religious faith. The antireligiosity also existed among the ancient Greeks and Romans. To name some: Lucretius, Epicurus, Democritus, Protagoras, Carneades, who were skeptical about religiosity. In early Islamic civilization, secularism was evidenced among many freethinkers such as Muhammad al-Warraq and Muhammad al-Razi in the ninth century, Omar Khayyam in the eleventh century, and Averroes in the twelfth century. What we see is that the evidences of various forms of freethinkers such as agnostics, atheists, naturalists, and secularists were widespread in the past. But now, in the twenty-first century, more people are leaving religion than embracing it (ibid.).

No religious affiliation is now the world's third-biggest "faith" after Christianity and Islam. Christianity is the largest faith with 2.2 billion adherents, or 31.5 percent of the world's population. There are about 1.6 billion Muslims around the world, or 23 percent of the global population. People with no religious affiliation now make up the third-largest global group in a new study of the world's faiths—coming after Christians and Muslims but just before Hindus. The study, based on extensive data for the year 2010, also showed that Islam and Hinduism are the faiths most likely to expand in the future, while Judaism has the weakest growth prospects. It showed Christianity is the most evenly spread religion, present in all regions of the world, while Hinduism is the least global religion with 94 percent of its population in one country, India.

Overall, 84 percent of the world's inhabitants, identify with a religion, according to the study titled "The Global Religious Landscape," issued by the Pew Forum on Religion and Public Life. The "unaffiliated" category covers all those who profess no religion, from atheists and agnostics to people with spiritual beliefs but no link to any established faith. "Belief in God or a higher power is shared by 7 percent of unaffiliated Chinese adults, 30 percent of unaffiliated French adults, and 68 percent of unaffiliated US adults," it said.

Pew Forum demographer Conrad Hackett said that the 2,500 censuses, surveys, and population registers were used to compile the report, but the data did not allow a further breakdown to estimate the world population of atheists and agnostics.

"It's not the kind of data that's available for every country," he said. "A census will typically ask what your religion is, and you can identify a number of particular affiliations or no religion."

An age breakdown showed that Muslims had the lowest median age at twenty-three years, compared to twenty-eight for the whole world population. The median age highlights the population bulge at the point where half the population is above and half below that number. Muslims are going to grow as a share of the world's population, and an important part of that is this young age structure. By contrast, Judaism, which has fourteen million adherents or 0.2 percent of the world population, has the highest median age at thirty-six, meaning its growth prospects are weakest.

Among more than a billion unaffiliated people around the world, 62 percent live in China alone, and they make up 52.2 percent of the Chinese population. Japan comes next with the second largest unaffiliated population in the world with seventy-two million, or 57 percent of the national population. After that comes the United States, where 16.4 percent of all Americans said they have no link to an established faith. The world's Hindu population is concentrated mostly in India, Nepal, and Bangladesh. Half of the world's Buddhists live in China, followed far behind by Thailand at 13.2 percent of the world Buddhist population and Japan with 9.4 percent. The study found that about 405 million people, or about 6 percent of the world population, followed folk religions such as those found in Africa and China or among Native American and Australian aboriginal peoples. Another fifty-eight million, or nearly 1 percent of the world population, belonged to "other religions," including Baha'i, Taoism, Jainism, Shintoism, Sikhism, Tenrikyo, Wicca, and Zoroastrianism. Most were in the Asia-Pacific region.[82]

The world of twenty-first century is very different from earlier times. We cannot go back to the time of the first century to understand events of occurrences in the same way that made sense in the worldview of an earlier generation. We do not live in a world where the sun stands still or where our god walks in the garden in the cool of the day or where voices speak to us out of flaming bushes. We see things differently from people of earlier generations. This is true for philosophers, historians, scientists, and theologians. We no longer accept that there is a religious explanation of natural events and processes. We understand a great deal about life, about laws of physics, about cosmos and of the atom, about the origin of the universe and the movement of the stars, about the space and time, about the evolution of life-forms and the earth's geologic formations, about the forces and matter, and about the cause and effect. We no longer use religion as an explanation to fill the gaps of our knowledge of things we do not yet understand. We do not try to harmonize the biblical seven days of creation with what we have learned from astrophysics about the structure and sequence of the origins of our universe. When we have questions about our physical world, we rely on the principles of science, whether the subject is matter, forces, energy, or the universe. Whatever we mean by the truth of the creation story, we accept the premise that is not intended to be a process description of an event in time. Each generation receives the truth of the previous generation wrapped in the concepts and framed out in the language and style that make sense to that generation. It is important to identify the central truth in religions, which must be discovered anew by each generation, and that each generation must reformulate of itself and in its own words.[83]

As children, we were raised as believers of the religion our parents follow. Whatever our parents told us, we believed. Upbringing, tradition, and, ultimately, fear seem to keep us locked into blind faith with no allowance of questioning that faith. We were not encouraged to think freely or question; in fact, it was just the opposite. That means don't question God, just accept it, do the rituals, and be like everybody else. This whole notion of nonquestioning servitude seems to have been effectively controlling billions of human pets for more than two thousand years. What a stupidity. I wonder how so many people can blindly accept illogical, irrational absurdities and believe them as true. The matter of fact is that our parents were never told by their parents that their inherited faith on religion was not the truth, and they passed it down to their children.

Similarly, we pass on the same religious message to our children. At some point in life, many of us start to critically evaluate what we were taught about the religion and what to believe. When we read the holy books, we begin to see the contradictions, the cruelty, and the mythology and then wonder how so many people put their heads in it rather than think of something logical, rational, and pragmatic way of life that can do better for us. But one and a half billion people on earth think that there is something fundamentally flawed with all the world's major religions. They dictate us to believe in an angry, jealous, self-proud, judgmental, all-powerful, loving god that demands worship, keeps monitoring everything we do, allows what we do, but will punish us for exercising our rights to question its nonsense contradictions. So being an adult, some of us develop a sensible, logical, and pragmatic attitude by rejecting religious stupidity. I hope readers will agree that knowledge and skepticism is a learned virtue. Religion is a complex, nonsensical ideology promulgated through fear and awe; it is a spiritually closed box with no window to see the light of knowledge of the rational world. So mankind's philosophy of life is humanity—live to do the most good and the least evil. We don't need holy books, priests, prophets, saints, monks, cardinals, goddesses or gods, and nonexistent crucified saviors.

Instead, humanity is to spread, at a higher rate, among world population. Only then will the source of our greatest fears and awe and superstitions burn into ashes where it belongs—that is, religion.[84]

The purpose of religion is to guide human life, to help us understand our needs and desires, and to establish guidelines for individually and socially acceptable ways to meet them. When these necessary and sufficient conditions are not met, the importance of religion falls apart. What we experience in our lives is that religion is dictating what our needs and desires should be.

Many people are happy and fulfilled following a religious creed. They like having a moral code, familiar rituals, a tradition, and a community that shares their values. But religion is not the only institution that provides those things. While some feel restricted by religious teachings that contradict modern metaphysical and ethical ideals, many others are frustrated by religion's illogical, orthodox, unethical, and, of course, monotonous services that bear no relationship to their actual lives when they search for meaningful, intellectual stimulation and vibrancy in life. On the contrary, nonreligiosity (such as atheism, agnosticism, secularism, and, in a broad sense, humanism) provides a better philosophy and guidelines for mankind to meet their needs, demands, and desires. Let me take an important issue—morality. I have discussed in some detail in the "Does the morality depend on religion?" section.

Let me refer to the research findings on morality. Phil Zuckerman, in his book *Society without God: What the Least Religious Nations Can Tell Us about Contentment*, found interesting results. Societies with high rates of happiness, stability, and social functioning are those with high rates of atheism. These societies have some of the lowest rates of violence crime in the world, some of the lowest rates of corruption, excellent education systems, strong economics, well-supported arts, free health care, egalitarian social politics, secular government, and more. However, this certainly doesn't suggest that high rates of atheism create happiness, stability, and high standards of social functioning. It's probably the other way around— happiness, stability, and high standard of social functioning tend to lead to high rates of atheism. This result, of course, compares with the societies with high rates of religious population. Zuckerman's results suggest that societies with high rates of atheism seem to be very moral societies with low rates of crime and corruption and a strong sense of social responsibility when compared to societies with high rates of religious population.[85]

A study—"Sex and Secularism: What Happens When You Leave Religion?"—conducted by organizational psychologist Darrel Ray in 2011 examined thousands of nonbelievers who once had religious beliefs. Research findings suggest that sexual behavior did not change when people become atheists. Religion and atheism do not affect people's sex lives—people engage in the same sexual acts whether they are believers of some religion or atheists. Interestingly, religion doesn't stop people from having kinds of sex that their religions forbid. It just makes them feel guilty about it. The idea that being nonreligious means abandoning one's self to amoralism, self-indulgence, and promiscuity is not true. Without religion, we don't need to borrow our ethics from thousand-year-old taboos introduced by some prophet or saint or savior who claimed that he speaks for God. People without religion can define their sexual ethics based on the core ethical values that human beings seem to have evolved with as a social species with good sense of kindness, justice and fairness, not doing harm, loyalty, the good social functioning, and so on. It is not necessary to say that having multiple sex partners is bad just because God says so. But it is important to decide whether having multiple sex partners is fair, whether the pleasure and joy it brings outweigh any possible harm it might do, whether it interferes with social stability, and so on. It is very much true that people without religion have sexual ethics. They have sexual ethics because they have tolerance, self-confidence, good judgment, and compassion.[86]

The feeling of knowing is very powerful. Our feeling of certainty always wins despite contrary evidence that should mitigate it. Our brain is very good at making up reasons to justify this feeling of certainty rather than following the evidence to the reasonable conclusions. Human beings have a very strong tendency not only to believe what they want to believe but also to seek and find evidence that confirms their beliefs. Yet many people have the power to overcome this and reject religion.

People reject religion because they could not answer the questions they were encountering. We can't think in a vacuum. Who we are and what we believe depends, to a large extent, on the influences that shape us. The bad experiences merely caused us to wake up from our dogmatic slumbers. They forced us into critical thinking about our inherited faith for the first time in our lives. People believe and doubt for a variety of reasons. Our experiences and our studies end our faith. We understand in the end that the skeptics are right. We open the holy books, and we find violence and the justifications of immoral acts rather than find wisdom.

We find no intellectualism but backward thinking. We find oppression, disparity, inequality, and suppression. Our prayers returned to us void. Living with the people around us, we find that we simply like the people for their attributes and characters, not because they believe something about religion. We need the people and their humane attributes, not the religion and its dogma. It is always right to question all that we believe. Clifford challenges religions by saying, "No religious belief system is capable of meeting the high standards required for belief, so no reasonable person should accept any religious belief system" (John and Lotus 2012). We must have the right to question what we believe. When we believe something unworthy, unethical, illogical, and unreasonable, we lose or weaken our power of self-control and self-confidence. The more we can question on anything we believe, the more it is good for the society.

How can one believe that someone heard a voice from the sky and claimed that it was from God commanding him to sacrifice his son by slaughtering or killing people or raping women, looting others' property (herds, livestock) or making men and women as slaves, robbing and plundering neighboring people, or doing utterly barbaric acts? Who, as a rational human being, is going to believe that the voice came from a perfectly good god? When we read such god in the holy books, our moral consciences don't accept it. We can't believe that it was commanded by a perfectly good god. We think that either the holy books are not words of God or such a god does not exist. All the arguments show that morality does not come from God. We don't need holy books or God to be good human beings. We can have a good, moral, and ethical life without God and religion. We are better off by enduring the principle of humanism with no religion.

Nonreligious people describe themselves in various ways, and those variations are likely to reflect differences in meaning and emphasis, although there are considerable overlaps. Nonbelievers (those who do not believe and practice any of the organized religions) do have some beliefs, although not religious ones. For example, they believe that moral feelings are social in origin, based on treating others as they would wish to be treated; it is, in fact, the "golden rule" that antedates all the major world religions. Among many beliefs, the most common ones are described below. While those nonreligious beliefs began in the long past, open and strong denial of religion became pronounced only during the nineteenth and twentieth centuries. During these periods, organizations have been built to represent the interests of the nonreligious people. The following are main terms that are commonly used to describe nonreligious peoples and their understandings that led to nonreligious beliefs, organizations, and movements.

Agnosticism

Agnosticism is the belief system that says that knowing of the existence or nonexistence of God is impossible. The people who believe in agnosticism are agnostic; in normal usage, it means "don't know" or having an open mind about a religious belief, especially the existence of God. It can also mean something much that nothing is known, or can possibly be known, about God or supernatural phenomena. The agnostic holds that human knowledge is limited to the natural world, that the mind is incapable of knowledge of the supernatural. Agnosticism can take a middle ground between theism and atheism. Agnosticism may simply be the state of not knowing whether any gods exist or not, but people can take this position for different reasons and apply it in different ways. According to philosopher William L. Rowe, agnosticism is the view that human reason is incapable of providing sufficient rational grounds to justify either the belief that God exists or the belief that God does not exist.

Most agnostics, however, may be categorized depending on how their beliefs work out in their day-to-day life practices, meaning whether they are more atheistic or theistic. Agnostics may live and act as if there is no god and no religion is correct, but they do not conform to be an atheist because of the expression of certainty. On the other hand, someone may consider them spiritual but not religious or perhaps even nominally following a religion but are identified as agnostics to convey an honest doubt about the reality of it all. The earliest professed agnostic was Protagoras, although the term itself (from the Greek *agnosis*, meaning "without knowledge") was not coined in English until the 1880s by T. H. Huxley. However, it was the philosopher David Hume who laid the foundations for modern agnosticism when he asserted that any meaningful statement about the universe is always qualified by some degree of doubt. Some of the most important agnostic philosophers are Protagoras, T. H. Huxley, Robert Ingersoll, and Bertrand Russell, and many world-renowned people have also been confessed agnostics, including Charles Darwin, Albert Einstein, Milton Friedman, Carl Sagan, Friedrich Nietzsche, Mark Twain, Richard Dawkins, and many others.

Huxley was the one who coined the terms *agnostic* and *agnosticism* to sum up his own position on metaphysics. His agnosticism was a response to the clerical intolerance of the 1860s as it tried to suppress scientific discoveries that appeared to clash with scripture. About the origin of the name agnostic, Huxley gave the following statement: "When I reached intellectual maturity and began to ask myself whether I was an atheist, a theist, or a pantheist; a materialist or an idealist; Christian or a freethinker; I found that the more I learned and reflected, the less ready was the answer; until, at last, I came to the conclusion that I had neither art nor part with any of these denominations, except the last. The one thing in which most of freethinkers agreed was the one thing in which I differed from them. They were quite sure they had attained a certain 'gnosis'—had, more or less successfully, solved the problem of existence; while I was quite sure I had not, and had a pretty strong conviction that the problem was insoluble. And, with Hume and Kant on my side, I could not think myself presumptuous in holding fast by that opinion. So I took thought, and invented what I conceived to be the appropriate title of 'agnostic.' It came into my head as suggestively antithetic to the 'gnostic' of Church history, who professed to know so much about the very things of which I was ignorant . . . To my great satisfaction the term took."

There are several types of agnosticism according to the differences of views of the existence of a god or gods or supernatural intelligence.

Agnostic theism. This is the view (also called religious agnosticism) of those who do not claim to know of the existence of a god or gods but still believe in such an existence.

Agnostic atheism. This is the view of those who claim not to know of the existence or nonexistence of a god or gods but do not believe in them.

Strong agnosticism. This is the view (also called hard agnosticism, closed agnosticism, strict agnosticism, absolute agnosticism, or epistemological agnosticism) that the question of the existence or nonexistence of a god or gods is unknowable by reason of our natural inability to verify any experience with anything but another subjective experience.

Mild agnosticism. This is the view (also called weak agnosticism, soft agnosticism, open agnosticism, empirical agnosticism, or temporal agnosticism) that the existence or nonexistence of a god or gods is currently unknown but is not necessarily unknowable; therefore, one will withhold judgment until more evidence becomes available.

Pragmatic agnosticism. This is the view that there is no proof of either the existence or nonexistence of a god or gods.

Apathetic agnosticism. This is the view that there is no proof of either the existence or nonexistence of a god or gods. However, since any god or gods that may exist appear to be unimportant and unconcerned for the universe or the welfare of its inhabitants, the question is only academic.

Ignosticism. It is the assertion that a coherent definition of God must be put forward before the question of the existence of God can be meaningfully discussed. If the chosen definition is not coherent (that is, not empirically testable), the ignostic holds the view that the existence of God is meaningless. The term *ignosticism* was coined by Reform Jewish Rabbi Sherwin Wine. It should be noted that some philosophers (like A. J. Ayer, Theodore Drange, and others) see ignosticism as different from atheism and agnosticism on the grounds that atheism and agnosticism still accept the existence of God as a meaningful proposition that can be judged to be false (that is, atheism) or still inconclusive (that is, agnosticism).

Ingersoll, known as "the Great Agnostic" and a strong supporter of free thought (the philosophical viewpoint that holds that beliefs should be formed on the basis of science and logic but not be influenced by emotion, authority, tradition, or dogma), justified the agnostic view point, which was summed up in his 1986 lecture "Why I Am an Agnostic." Charles Darwin was raised in a religious environment and was an Anglican clergyman. At some point, he eventually doubted parts of his faith but continued to help in church affairs, even while avoiding church attendance. He stated that it would be absurd to doubt that a man might be an ardent theist and an evolutionist at the same time. Although his religious views were not revealing, he wrote, "I have never been an atheist in the sense of denying the existence of a god. I think that generally, an agnostic would be the most correct description of my state of mind."

Bertrand Russell's "Why I Am Not a Christian" and "Am I an Atheist or an Agnostic?" are considered classic statements of agnosticism. He was very careful to distinguish between his atheism with regard to god concepts and his agnosticism with regard to superhuman intelligence. Although he generally considered himself an agnostic in a purely philosophical context, he emphasized that the label "atheist" represents a more accurate understanding of his views in a popular context, meaning he was more an atheist than an agnostic. He said in his lecture, in 1939, on *the existence and nature of God*, in which he characterized himself as an atheist. He also considered himself as agnostic as has been found in the 1947 pamphlet *Am I an Atheist or an Agnostic?* He said, "As a philosopher, if I were speaking to a purely philosophic audience, I should say that I ought to describe myself as an agnostic, because I do not think that there is a conclusive argument by which one can prove that there is not a God." About the evidence that could convince him that God exists, he said, "I think that if I heard a voice from the sky predicting all that was going to happen to me during the next twenty-four hours, including events that would have seemed highly improbable, and if all these events then produced to happen, I might perhaps be convinced at least of the existence of some superhuman intelligence." However, his final thought is that there is no reason to believe any of the dogmas of traditional theology and that there is any reason to wish that they are true. Human beings insofar are not subject to natural forces and, therefore, are free to work out their own destinies. Both the responsibility and the opportunity are his but not of any divine power.

Atheism

The term *atheism* comes from the Greek word *atheos*, meaning "godless." *Atheos* is derived from *a*, meaning "without," and *theos*, meaning "deity." An atheist has no religious belief. An atheist does not believe in a god or gods or other supernatural entities. They reject a belief in the existence of God or gods and those who simply choose to live without God or gods and often disbelieve in the soul, afterlife, and other beliefs arising from god-based religions. An atheist has no specific belief system. They accept only what is scientifically verifiable. Since god concepts are unverifiable, they do not accept them.

Atheism is commonly divided into two groups—strong atheism and weak atheism. This distinction arises mainly from their understanding on the existence of gods. Weak atheism simply means the absence of belief in any gods. A weak atheist is someone who lacks theism and who does not happen to believe in the existence of any gods. This is also sometimes called agnostic atheism because most people who self-consciously lack belief in gods tend to do so for agnostic reasons. Strong atheism involves not only believing but also denying the existence of any god or gods at all. Strong atheism is sometimes called gnostic atheism because people who claim to have the knowledge that a certain god or gods do not or cannot exist. Philosopher Ludwig Feuerbach and psychoanalyst Sigmund Freud have argued that God and religious beliefs are human inventions, created to fulfill psychological and emotional needs. Many others like Karl Marx, Friedrich Engels, and Mikhail Bakunin argued that belief in God and religion are social functions used by those in power to oppress the working class and said that the idea of God implies the abandonment of human reason and justice. It is like a reversal of Voltaire's principle, "If God did not exist, it would be necessary to invent him," but writing instead, "If God really existed, it would be necessary to abolish him."

Atheism is somewhat acceptable within some religious and spiritual belief systems, including Hinduism, Jainism, Buddhism, Syntheism, realism, and neo-pagan movements such as Wicca. Astika schools in Hinduism promote atheism to be a valid path to Moesha. Jainism believes the universe is eternal and has no need for God or a creator deity. However, Tirthankaras are admired deeply that can transcend space and time and have more power than the god Indra. Secular Buddhism does not conform to belief in gods. Early Buddhism was atheistic, as there was no mention of gods in Gautama Buddha's path of enlightenment. It is important to mention that atheistic schools were found in early Indian philosophy and have existed from the times of the Vedic religion. Among the six orthodox schools of Hindu philosophy, Samkhya (the oldest philosophical school of thought) does not accept God, and the early Mimamsa also rejected the notion of God. History suggests that the thoroughly materialistic and antitheistic philosophical Carvaka (also called *Nastika* or *Lokaiata*) school that was established in India around the sixth century BCE is probably the most explicitly atheistic school of philosophy in India, which is similar to the Greek Cyrenaica School. This branch of Indian philosophy is classified as heterodox because of its rejection of the authority of Vedas.

Although it is not considered part of the six orthodox schools of Hinduism, however, it is contextually important to cite as evidence of a materialistic movement within Hinduism. The Sufi Jalal al-Din Rumi (who once was a teacher of religion, science, and mysticism) was a poet and a philosopher. He was regarded as the greatest literary figure in the Islamic world. He once proclaimed, "As to my creed, I am neither a Jew nor Zoroastrian, not even a Muslim as the term is generally understood." When he was accused of being an atheist and heretic, he agreed (Burns 2006).

Plato (in his *Euthyphro* dilemma) argued that the role of the gods in determining right from wrong is either unnecessary or arbitrary. There has been a persistent issue in the political and in the philosophical debate on the argument that morality must be derived from God and that it cannot exist without a wise creator. Theists believe that moral precepts such as "Murder is wrong" are seen as divine laws, which require a divine lawmaker and judge. However, atheists argue that morality does not depend on a lawmaker in the same way that laws do. Friedrich Nietzsche believed in a morality that is independent of theistic belief, and he stated that morality based upon God "has truth only if God is truth—it stands or falls with faith in God." There exist normative ethical systems that do not require principles and rules that are believed to be given by a deity of God. For example, some of the ethical systems include virtue ethics, social contract, Kantian ethics, utilitarianism, and objectivism. According to Sam Harris, not only is moral prescription an issue to be explored by philosophy, but one can also meaningfully practice a science of morality. Any such scientific system must respond to the criticism that is embodied in the naturalistic fallacy.

Philosophers Susan Neiman and Julian Baggini asserted that behaving ethically in the belief of divine mandate only is not the true ethical behavior but merely a blind obedience. The contemporary British political philosopher Martin Cohen brought forward the historical example of biblical injunctions, which favored torture and slavery in the name of religion and how religious injunctions follow political and social customs, rather than vice versa. Cohen further extends this argument in more detail in the *Political Philosophy from Plato to Mao*, where he argued that the Quran played a role in perpetuating social codes from the early seventh century despite changes in secular society. Some prominent atheists such as Christopher Hitchens, Daniel Dennett, Sam Harris, Richard Dawkins and thinkers like Bertrand Russell, Robert G. Ingersoll, Voltaire, and José Saramago have criticized religions by citing harmful aspects of religious practices, rituals, and doctrines. The German political theorist and sociologist Karl Marx criticized religion by saying, "The sigh of the oppressed creature, the heart of a heartless world, and the soul of soulless conditions. It is the opium of people."

The 2012 survey on the *Global Index of Religiosity and Atheism* examined how people viewed themselves as "a religious person, not a religious person or a convinced atheist." The results show that the top ten countries with people who viewed themselves as "convinced atheists" are China (47 percent), Japan (31 percent), the Czech Republic (30 percent), France (29 percent), South Korea (15 percent), Germany (15 percent), Netherlands (14 percent), Austria (10 percent), Iceland (10 percent), Australia (10 percent), and the Republic of Ireland (10 percent).[87]

Freethinkers

Freethinkers are the people who reject authority in matters of belief, especially religious beliefs. It was a very popular term in the nineteenth century and is still used in different languages in some European countries by nonreligious organizations to describe themselves as humanists.

They form opinions about religion on the basis of reason, independent of tradition, authority, or established belief. Freethinkers include atheists, agnostics, and rationalists. Freethinkers never conform to everything in the scriptures and religious books and any prophets or messiah. They think that revelation and blind faith are invalid and, thus, no guarantee of truth. The concept of freethinking refers to the process of making decisions and arriving at beliefs without relying solely upon tradition, dogma, or the opinions of authorities. Usually, the context of this is only in religion, although a person can be a freethinker in other areas as well. In place of tradition or dogma, freethinkers insist upon using reason, logic, and evidence as the bases for forming reasonable and justified beliefs. Superstition is rejected in favor of science. Most freethinkers are also atheists or agnostics, although that is not imperative. It is possible for someone to be an atheist without being a freethinker or to be a freethinker without being an atheist. Like humanists, freethinkers believe that moral values follow on from human nature and experience, rather than depend on a god or any supernatural agency.

Rationalism

Rationalism derives from the idea that accepts the supremacy of reason, as opposed to blind faith, and aims at establishing a system of philosophy, values, and ethics that are verifiable by experience, independent of all arbitrary assumptions or authority. The principal doctrine of rationalism holds that the source of knowledge is reason and logic. Thus, rationalism is contrasted with the idea that faith, revelation, and religion are also valid sources of knowledge and verification. Rationalists, in this context, prioritize the use of reason and consider reason as being crucial in investigating and understanding the world, and they reject religion on the grounds that it is unreasonable. Rationalism is in contradistinction to fideism, a position that relies on or advocates faith in some degree.

Skepticism

Skepticism has a long history dating back to ancient Greece when Socrates experienced, observed, and said, "All I know is that I know nothing." Modern skepticism is embodied in the scientific method that involves gathering data to formulate and test naturalistic explanations for natural phenomena. However, any claim becomes factual when it is confirmed to such an extent that it would be reasonable to offer temporary agreement. But all facts in science are provisional and subject to challenge and proof, and, therefore, skepticism is a method leading to provisional conclusions. When some claim to have been tested and failed by the tests, we can provisionally conclude that they are false. Other claims have been tested and found that the results are inconclusive, so we must continue formulating and testing hypotheses and theories until we can reach a provisional conclusion. The key to skepticism is to continuously and vigorously apply the methods of science to prove hypotheses and theories.

Philosophical skepticism is very critical, which systematically questions the notion that absolute knowledge and certainty are possible. However, it is important to note that philosophical skepticism is opposed to philosophical dogmatism, which emphasizes the fact that a certain set of positive statements are authoritative, absolutely certain, and true. Philosophical skepticism should be distinguished from ordinary skepticism, in which doubts are raised against certain beliefs because the evidence for the particular beliefs is weak or lacking. Ordinary skeptics don't take things on trust, but they must see the evidence before believing. Ordinary skeptics doubt the miraculous claims of religions, the claims of alien abductions, the claims of psychoanalysis, etc. But they do not necessarily doubt that certainty or knowledge is possible. Nor do they doubt these things because of systematic arguments that undermine all knowledge claims.

Naturalism

Naturalism is a metaphysical concept that holds the view that all phenomena can be explained scientifically in terms of natural causes and laws as opposed to the supernatural. Naturalism emphasizes that the universe is a vast machine or organism without any general purpose. Naturalism neither denies nor affirms the existence of God, either as transcendent or immanent. However, naturalism makes God an unnecessary hypothesis and essentially superfluous to scientific investigation. Naturalism entails the nonexistence of all supernatural beings, including the theistic god. Naturalists not only hold that the view that evidence for the supernatural has not been convincingly demonstrated but also think that the belief in the supernatural has led to a great deal of misery for humanity and, thus, needs to be rejected and replaced with critical inquiry, accountability, and science.

Secularism

Secularism is the belief that religion should be a private, personal, voluntary affair that does not impose upon other people. It ensures that religions are treated fairly and that no bias exists for a particular religion and also that nonreligious people are treated with equal respect. It is the only democratic way to prosper in a globalized world where populations are free to choose their own varied religions and lead their lives as good democratic citizens. *Secular* means "without religion" and that nonreligious people lead secular lives. The theory of secularization is that as society advances in modernity, religion retreats. The most likely fact is that intellectual and scientific developments undermine the spiritual, supernatural, superstitious, and paranormal ideas. As such, scientific development and intellectual endeavors weaken religion because it relies on supernatural dogmas for legitimacy. With the advancement of science, eminent scientists, sociologists, philosophers have realized that religion may be in a permanent decline.

With the advancement of human intelligence, education and science cause the demise of religion, as these influence people to be less religious. Also, the concept of human rights has identified many religious practices as barbaric and immoral, such as gender equality, racism, and prejudice against homosexuality, which wins over religious dogmas. Many eminent men played significant roles in favor of secularism against religion—for example, philosopher Friedrich Nietzsche, the sociologists Max Weber and Karl Marx, and the psychologist Sigmund Freud, activist humanist Prof. Paul Kurtz, and scientist Prof. Richard Dawkins. So one may argue that religion becomes hollow and will survive for a limited time until its active membership is reduced to a minimum.[88]

Pantheism

A person who follows the religious doctrine of pantheism believes that God is all around us throughout the whole universe. Pantheism implies a lack of separation between people, things, and God but rather sees everything as being interconnected. More rarely, *pantheism* refers to a belief in all gods from all religions or a tolerance for those beliefs. In Greek, *pan* means "all," and *theos* means "god."[89] According to pantheism, there is no real difference between "good" and "evil" because all is "god" in the end. In the pantheistic view, humans are "God" as well because they are part of the universe.

Science, Religion, and Philosophy

How is it that hardly any major religion has looked at science and concluded, "This is better than we thought? The Universe is much bigger than our prophets said, grander, more subtle, more elegant?" Instead they say, "No, no, no! My god is a little god, and I want him to stay that way." A religion, old or new, that stressed the magnificence of the Universe as revealed by modern science might be able to draw forth reserves of reverence and awe hardly tapped by the conventional faiths.

—Carl Sagan

Learn from yesterday, live for today, hope for tomorrow. The important thing is to not stop questioning.

—Albert Einstein

An Overview

The challenge of our time is to intelligently coordinate and unify the realms of science, philosophy, and religion so that we may move toward a greater comprehension of total cosmic reality, which can benefit individual lives as well as our entire civilization. Therefore, we need a balanced understanding of the true meaning and value of our individual, interpersonal, and social lives, as well as our place on earth and in the cosmos—a viewpoint that wisely integrates science, philosophy, and religion. We need science to enrich our knowledge and understanding about the laws of energy, matter, and our physical environment so that we may apply that knowledge to live better in our material world. We need philosophy to teach and help us how to think more clearly and to reasonably discern the universe of things, meanings, and values. Bertrand Russell mentioned (in his book *Wisdom of the West*) that the branches of knowledge have borders beyond which there are unknown and that when one passes from known (i.e., science) into unknown, there is speculation. This speculative activity is exploration, which is, among other things, what philosophy is. He further said that there are two ways to know the unknown—one is to accept what people say that they know on the basis of books, mysteries, or other sources, and the other way is to go out and look for yourself, and this is the way of science and philosophy. We need religion to guide us to lead our lives, to know our true relationship with God and one another and spirit values so that we may become increasingly God-conscious, to live in harmony with one another, and to make progress toward our mutual divine destiny. However, one can argue that the philosophy of religion is neither ceremony nor ritual nor going to the church, temple, mosque, or synagogue, but an inner experience that finds God everywhere in nature. Here is the question, which kind of religion do we need for mankind?

Combining the Three Themes

While the knowledge of humankind is divided into areas of science, philosophy, and religion, they themselves are further subdivided into a number of various specializations, schools of thought, and conflicting doctrines. The different specializations of science, philosophy, and religion were not the same in the past as it is today. For example, Plato, being a philosopher, dealt with religious matters, and Pythagoras was a mathematician and a philosopher who explored the nature of the universe and was revered as a religious person as well. Although not all people were religious and many thinkers in the past rejected spiritual notions, it was much easier to be religious, philosophical, and scientific without contradiction. At a later time during the Renaissance, a man like Francis Bacon was of the opinion of total integration of science, philosophy, and religion. He helped to lay the foundation of empirical sciences, being a mystic, Rosicrucian, and Freemason. Other great thinkers during the same time such as Galileo and Copernicus significantly contributed to the advancement of science and remained devotedly religious. Sir Isaac Newton integrated all the scientific knowledge and developed a coherent mathematical framework and spent time thinking about spiritual and religious matters at the same time. He stated that science was not something that replaces God but rather an investigation into the logical beauty of creation.

In the eighteenth century, around the time of Enlightenment, a split between science and religion started to occur because scientific thinking has evolved. Influential thinkers were pursuing and advancing philosophical and scientific reasons aimed at establishing authoritative systems of ethics and aesthetics. They were advocating that reason and science would be the only means by which people could gain better knowledge and understanding of the universe and existence. The combined effect of Newtonian physics that explained the heavens and Darwin's evolutionary ideas that explained the emergence of all living things on earth seemed to disregard the necessity of God as the creator. In later time, many scientists were antagonistic to religion and the idea of God. Albert Einstein talked about a sense of the mystical as the source of all true art and science. He also said, "Science without religion is lame, religion without science is blind" and "God does not play dice." By these statements, however, one should not be convinced that he believed in a personal god without having a more detailed definition concerning the exact relationship between science and religion (as I have discussed earlier). Much later, the celebrated cosmologist Stephen Hawking, in his books *A Brief History of Time* and *The Grand Design*, has described how to eliminate God as the primary cause of existence. An evolutionary biologist and a strong proponent of Darwinism, Richard Dawkins promoted a materialistic and atheistic viewpoint. These are only few names among many others.

There is a real antagonistic notion toward areas of science by the religious community. For example, in the area of evolution, genetic research concerning the origin of the universe and the earth, they are very much opposed to it. They think that the rise and power of scientific rationalism are closely related to secular humanism, and, thus, it is a threat to their belief system. We can pose a question whether it is possible that secular humanists and religious fundamentalists will find a common ground for reconciliation. The reconciliation of science, religion, and philosophy is plausible only when the central truth prevails in all world religions, in scientific discoveries, and in philosophical matters.

Quantum mechanics explains the behavior of matter and its interactions with energy on the scale of atoms and subatomic particles. Matter can go from one spot to another without moving through the intervening space. The two subatomic particles existing in the same point in space and time will become coupled to one another in such a way that they become like one entity. Scientists have discovered that the entire universe is actually a series of probabilities. The results from the realm of quantum mechanics are challenging existing assumptions about the laws of the physical world (i.e., matter, energy, and motion), and these, in turn, are confirming mystical ideas from the world of religion that have existed for thousands of years. The attribute and indefinite nature of the subatomic particles have compelled us to reconsider our ideas about the nature of world and reality. Results from quantum physics state that the entire universe and everything in it are likewise entangled and, therefore, consist of a single unified "oneness." Therefore, science is converging toward timeless ideas from religion concerning the oneness of all things and the oneness of God. It is interesting to note that the terms *God* and the *all* that is the universe are often used interchangeably in religious scriptures. Now this idea of God's oneness is perhaps the central tenet of the so-called monotheistic religions—Judaism, Christianity, and Islam. Moreover, the oneness of God is also attested to in other religions and is really a universal truth behind all the world's great faith traditions.

Quantum physics tells us about behavior of particles at the subatomic level, explaining the unity and inseparability of all existence, which, in turn, allows us to justify the validity of the claims that we are part of God and the universe. Looking through this light, we seem to experience the oneness that quantum physics is suggesting the true nature of things, all-encompassing oneness that may be called god. This idea has close resemblances to Spinoza's god. God is an absolutely infinite being—that is, substance consisting of infinite attributes, each of which expresses eternal and infinite essence. Spinoza believed that everything that exists is God, but he did not hold the converse view that God is no more than the sum of what exists. I have discussed Spinoza's god in the previous section.

We may argue that the cutting edge of science tends to confirm timeless mystical truths that exist at the core of the world's religions. More truth will be revealed in future with the continuing progress of science, resulting in well-founded convergence between science and religion. But we must think that this is not about the substantiation through science of those aspects of religion (outer mysteries or exoteric traditions such as the rules, rituals, regulations, fairy tales, and fantasies). Rather, it is the inner mysteries, the esoteric truths, and the mystical heart that exist in all the world's religions. Only within the inner mysteries can one discover a set of beliefs that are common to the esoteric traditions of all the world's religions. It is these inner mysteries that scientific discoveries are likely to confirm and that only the inner truths of religion may help science reconcile with religion.

What is the main tenet in terms of philosophy regarding the esoteric faith of religion that can be put into context? Two branches of philosophy may be worth noting: ontology (what is the nature of things?) and epistemology (how do we know of things that we think we know?). Ontology can be divided into two opposite viewpoints—materialism and idealism. Materialism supposes that the nature of existence is based on matter, whereas the main tenet of idealism is the belief that all existence is consciousness. It is idealism that is convergent with ideas from the world of religion and mysticism. Consciousness is related to the descriptions of God's attributes, since God is universally described as immanent and within us, which is in all the world's great faiths. Thus, we can say that consciousness is also immanent and within us. God is also universally described as transcendent—that is, independent of and beyond the realm of matter. In this sense, consciousness is also transcendent and somehow above and separate from the material world. Materialist philosophers and neuroscientists have been working on this area.[90]

Prof. Sankarshan Achraya[91] has presented substantial new insights about epistemic truths to help resolve current problems facing humanity worldwide. He discussed on a coherent, unified philosophy about how humans have universally formed beliefs to govern themselves and how this philosophy could help resolve current problems. "Humanity now begs to answer a fundamental question of how we can govern ourselves," he said. I will try to outline his views, as much as possible, about the unification of science, philosophy, and religion.

We, as humans, have formed beliefs about the set of unknown elements of the universe. Such beliefs are called probability beliefs in mathematics, science, and engineering. We have branded our beliefs about the elements in the unknown set as religion. In the probability theory, we measure the unknown based on some perceptions, which we can call beliefs in religious term, about the uncertain elements of the universe, which is a rationally defined doctrine or which we may call religion. So the probability theory is a kind of belief, and the formation of belief process about the unknown elements is common to all fields, including religion. However, this common process does not automatically produce a unifying philosophy because of the incoherence in beliefs about God across religions. The atheists, scientists, and those who do not associate with any religion do not accept God. The challenge here to articulate a unifying philosophy rests on presenting God in a way that will be acceptable to the people of all religions, as well as to the scientists, atheists, and nonreligious people.

To present God coherently, let us divide the set of all elements in the universe into two disjoint subsets. The first subset is composed of all elements of the universe that are known to humans; let's call this set as knowledge. The second subset comprises the rest of the elements of the universe; let us call this set as unknowable. Both the disjoint sets are not static in any sense; rather, they are evolving dynamically over time. Knowledge is expanding. Unknowability is shrinking, but it still remains infinite. It is logical and reasonable to say that the unifying philosophy presents the unknowable set as a universal god and the beliefs about the elements in this set as universal beliefs. We call the set of unknowable elements universal god because it forms the common basis of characterizing God in all the existing religions. The universal god is known as the set of unknowable elements. This is unlike the existing proclamation by many that God is unknown, Professor Sankarshan argued. God cannot be unknown.

A coherently rendered god should comprise the whole unknowable set of elements. It makes sense. Let us suppose that God is unknown (as believed by the followers of religions). Having said so, we see that an unknown god will simply be a subset of the unknowable set. But every religion also admits that God is almighty. This means God cannot be a mere subset of the unknowable. It is because such a subset will exclude some elements of the unknowable that are not within the reach of God. This is not likely the case. Therefore, God, as the unknown almighty, must be the entire unknowable set. The unknowable is, thus, a coherent rendition of a universal god that can be accepted by followers of religions, as well as the scientists, atheists, and nonbelievers of religions. It would be logically correct to say that the longing to reach a universal god in universal religion is called scientific research, perseverance, and tenacity.

It has been established that the fact that the ability to conform is inherent in human genes. The human gene is able to store observed facts as knowledge and then verify conformity of a new discovery or claim to the stored facts to determine the truth about the discovery or claim. The human gene is the repertoire of memory like epistemic logics. The gene mutates at birth. The repertoire of memory (e.g., epistemic logic), too, is mutated (not completely erased) at birth. We see that a newborn child responds through cries when the care is required. This is a genetic response to conform facts as necessary conditions for survival. As the child grows, the gene accumulates new knowledge. It is the gene's nature to retain epistemic element that is needed to conform, as fact, the conditions necessary for coexistence (cosurvival) of its carriers (the humans). Perhaps this is the reason humans have found commonly acceptable tenets for coexistence or cosurvival, which may be called religion. The similar analogy may be applied in modern science to the fact that science became a new religion in the context of the mathematical probability (as belief) about unknown elements of nature (uncertainties). It is found that the epistemic elements (facts), retained within the genes, may be dormant in some humans. Thus, these humans do not automatically know the commonly acceptable rules for coexistence. As a result, these humans may likely be subjected to harsher and uncommon rules, rituals, and social conditions to lead life. However, it is possible for some people to activate the dormant genes that have active epistemic elements (facts) in their genes. Having activated the dormant genes, they can conform, as fact, the truth about some commonly acceptable new rule for coexistence. Since the nature of the gene is to survive, any passive element within it is most likely to be activated when the gene's carrier (human) is faced with survival challenges, Sankarshan suggested.

We have been experiencing that existing world religions do not unify global population and do not bring prosperity and stability that every human inherently cherishes. Preaching that some existing religion is superior to others is not conducive to coexistence. Only the unifying philosophy of universal religion (not the existing religious beliefs) induces humans to persevere and produce to attain prosperity and stability for humans. Only this philosophy of world religion vis-à-vis humanity can bring harmony among global population. Promoting and establishing this philosophy globally can unify humanity and render all other religious beliefs superfluous.

Conceptualizing God as the set of elements unknown, we can remove confusions and ambiguities about God and knowledge. Seeking new knowledge is like discovering the elements of the unknowable through a process that is known as scientific research in modern thinking. Similarly, the traditional prayer to God is a process known as religious rules and rituals. Considering that scientific research and prayer are both intended to uncover the truth, we can bring them to the domain of knowledge that some elements are an unknowable set or are part(s) of a universal god. The process of searching for the truth is the aim of science. If the same idea is true for religion, only the esoteric element of religion is the search for the truth. Three things come into consideration: (1) the unknowable is God, (2) science is knowledge, and (3) belief about the unknowable is religion.

Such transparent and coherent definitions can remove the confusions about religions and science.

Like Professor Sankarshan, many of us acknowledge the fact that tenets of any religion should be based on enhancement of stability and prosperity of humankind. As we think that democracy is the best (though not ideal) form of governance accepted by humans as optimal, democracy should be the fundamental tenet of universal governance systems (say, universal religion). Similarly, amending constitutional rules of law through optimal discourse and vote within democracy is the other tenet. As Sankarshan pointed out, Gita in Hinduism, Bible in Christianity, Quran in Islam, and scripts in other religions were meant to be "guiding" rules of governance for mankind. But these scripts have never been amended to incorporate the latest human wisdom, including the democratic process of creation and amendment of the constitutional rules of law. They remain unchanged for centuries until today and, thus, remain dogmatic and sacrosanct. The countries that have unshackled their governance from such dogma have enhanced their prosperity and stability than those that remain shackled. Accepting universal religion does not mean abandonment of current beliefs of an individual. It is really a very conducive and healthy atmosphere to live in harmony among people, because no individual will ever feel anything wrong about accepting universal religion as in a conversion to a different religion. Moreover, universal religion gives a complete freedom to choose and to amend or refine the scriptures through rational arguments to synchronize with human wisdom, Professor Sankarshan concluded.[92]

The aim of science is to establish general rules that determine the reciprocal connection of objects and events in time and space. For rules or laws of nature, absolutely general validity is required for sure. As for religion, a doctrine that maintains only in the dark loses its effect on mankind with incalculable harm to human progress. Religious teachers must have the moral and mental strength to give up the doctrine of a personal god—that means to give up the source of fear and hope. To be noted here is that the priests, imams, and religious preachers hold the power to infiltrate the minds of weak-minded people. Instead, they should direct them to use those forces that are capable of cultivating the good, the true, and the beautiful in humanity itself. Religious teachers can accomplish this, and they will surely recognize with joy that true religion has been ennobled and made more profound by scientific knowledge as a result, Einstein suggested.

Scientific reasoning can aid religion to liberate mankind as far as possible from the bondage of egocentric cravings, desires, and fears, if this is the goal of religion. Science not only purifies the religious impulse of the dross of its anthropomorphism but also contributes to a religious spiritualization of our understanding of life, as Einstein explained. The further the spiritual evolution of mankind advances, the more certain that the path to genuine religiosity does not fall through the fear of life, the fear of death, and blind faith, but through striving after rational knowledge. In this sense, he believes that the priest must become a teacher if he wishes to do justice to his lofty educational mission. This idea of universal religion, in fact, is closely aligned with humanism. It is the time to bring harmony among mankind (people of all religions) by following the footsteps of humanism—a way of life that is being practiced by almost 15 percent of the world's seven billion people.

John Loftus, in his book *Why I Became an Atheist*, discussed in some length about the relationship between science and religion. He quoted David Eller, who wrote the following:

Not all people, even all religionists, are hostile to science. I have never seen a serious attack on atomic theory or quantum theory or gravitational theory or, on the whole sciences like meteorology or botany or germology. Religion is generally unconcerned with these sciences, because these sciences are unconcerned with the questions that concern religion. So, one of the main failures in the battle between science and religion is the tight focus on a very few bits of science and the generalization that they (mostly evolutionary theory and the sexual or reproductive sciences) are science. So, it is not a meaningful question to ask whether religion is compatible with science; it depends on which religion, which science, and what one means by compatible.

This is an interesting point to be looked at. Why do the religious people accept the results of science in many areas but object to some areas that contradict what ancient superstitious and orthodox preachers wrote in Holy Scriptures? To explain the unresolved areas between science and religion, John Loftus has reproduced arguments of anthropologist Ian Barbour, an eminent scholar on the subject who presented four ways of relating science and religion. They are *conflict, dialogue, independence,* and *integration.*

The fundamental conflict between science and religion arises from two themes—scientism and literalism of Holy Scriptures, which never reconciles. Nonbelievers think that scientism (also known as scientific materialism) scientific theories and methods are the principal means of knowledge and that matter is the fundamental reality of the universe. While the first one is based on epistemology, the second one is based on metaphysics, whereas scriptures' literalism takes a literal interpretation of the Holy Scriptures and sets the limit for science, meaning science cannot cross the limit of what holy books interpret. This is where science and religion come into direct conflict. The people who are educated and scientifically conscious reject the untrustworthy approach of Holy Scriptures' literalism. As Neil deGrasse Tyson argued, it is not seen that a successful prediction about the physical world has been inferred from the content of any religious scriptures. On the contrary, whenever anyone has used any holy scriptures or any religious documents to prove or make predictions about the physical world, it was found to be wrong. Here, prediction means an accurate statement about the untested behavior of objects or phenomena in the natural world (Neil deGrasse Tyson).

Dialogue takes position when science and religion go side by side on some aspect and contradict on some others. It is argued that religious beliefs interpret and correlate experience, whereas scientific theories interpret and correlate experimental evidence. It would not be appropriate to say that science explains and informs but that religion reveals and reforms. This viewpoint has been explained by Donald MacKay: Both science and theology give different kinds of explanations (with different methods and aims) about the same objects. Both explanations of the same event can be true and complete on their own levels. But the methods differ greatly. Compare how an artist, poet, theologian, or astronomer might view a sunset. They can all be correct from their perspective, even if they disagree with one another. There is no incompatibility in claiming that the formation of the universe, as we know, is the result of natural processes and that "the cosmos is God's creation." Each explanation is from a particular conceptual framework and can be true from the perspective of that framework.

Others have similar arguments. For example, Howard Van Till has argued that when scientists make statements concerning the origin, governance, or purpose of the cosmos, they are necessarily going outside the boundary of science and drawing from their religious and philosophical perspectives. Similarly, theologians do the same thing. They make statements about the geologic processes or thermodynamic phenomena or cosmic chronology. They are necessarily going outside the boundary of scriptural interpretation and getting into the domain of modern natural science (Howard Van Till). But this kind of dialogue between science and religion is unlikely to happen, because both parties have very different views, and they justify points of view with their own perspectives. Richard Dawkins thinks that it is not necessary to have a dialogue between science and religion and said that why we should not comment on God as scientists. A universe with a creative superintendent would be a very different kind of universe from one without. Why is that not a scientific matter? Science concerns itself with *how* questions, but only theology is equipped to answer *why* questions. The question *why* can't be answered. There are some genuine questions that are beyond the reach of science. But if science cannot answer some ultimate question, then what makes anybody to think that religion can? Richard Dawkins has asked this question. This is, of course, a good question. Take a look on religious perspective. Theologians have no evidence of any natural phenomena; their tenet is only belief, nothing more than that.

Science and religion stand on two different platforms, and they are independent of each other. They have contrasting arguments based on two different methods. Science is based on evidence and deals with nature, whereas religion is based upon belief and deals with God. The view of independence between science and religion may be cited by the idea of NOMA (nonoverlapping magisteria)—a concept brought by Stephen Jay Gould that states that science and religion are two mutually independent realms. More specifically, science occupies the empirical realm of fact and theory, whereas religion deals with ultimate meaning, purpose, and moral values (Loftus 2006).

According to Barbour, science and theology can be integrated with each other. Integration model suggests that science and theology may be combined to create a more coherent view of reality. This model was exemplified in the works of Pierre Teilhard de Chardin, who sought to integrate evolution, Christianity, redemption, and perfection, and saw all this fulfilled in his vision of the "Omega Point." Ian Barbour thought that the content of science and theology can be integrated in at least three ways. They are *natural theology*, *theology of nature*, and *systematic synthesis*.

Theologian Thomas Aquinas said that some of God's characteristics can be known only from revelation in scripture, but the existence of God can be known by reason alone. Darwin argued that adaptation can be explained by random variation and natural selection, but later, he revised the argument as saying that God did not design the particular details of individual species but designed the laws of evolutionary processes through which the species were formed, leaving the details to chance. So natural theology states the fact that understanding of nature can make us think and support theology as suggested by Thomas Aquinas.

Theology of nature does not start from science; rather, it starts from a religious tradition based on religious experience and historical revelation. However, theology of nature states that some traditional doctrines need to be reformulated in the light of current science. Here science and religion are considered to be relatively independent sources of ideas, but with some areas of overlap in their claims. So our understanding of the characteristics of nature is likely to affect our models of God's relation to nature. In contemporary views, nature is understood to be a dynamic evolutionary process characterized by both law and chance. This suggests that the natural order is ecological, interdependent, and multileveled. These characteristics can modify our thinking of the relationship between God and humanity to nonhuman nature, which, in turn, affects our attitudes toward nature. Biochemist and theologian Arthur Peacocke supported the harmonization of religious belief and scientific theories. He is willing to reformulate traditional beliefs in response to current science by discussing how chance and law work together in cosmology, quantum physics, nonequilibrium thermodynamics, and biological evolution.

Ian Barbour suggested that systematic synthesis is a systematic integration that can occur if both science and religion contribute to a coherent worldview elaborated in a comprehensive metaphysics. The process philosophy of Alfred North Whitehead and others supported a process theology that may be called process philosophy, which asserts that the universe is the creation of God in his entities. According to him (as for evolutionary thinkers), nature is a dynamic web of interconnected events characterized by novelty as well as order. Process thought states that the basic constituents of reality are not two kinds of enduring entity (i.e., mind/matter dualism) or one kind of enduring entity (materialism) but one kind of event with two aspects or phases. All integrated events have an inner and an outer reality, but these take very different forms at different levels. God elicits the self-creation of individual entities, thereby allowing for freedom and novelty, as well as order and structure (Barbour).

The origin of religious beliefs was prescientific. The people of those times had little or no understanding of the natural processes such as sunrise, rainfall, earthquake, thunderstorm, cyclone, and so on. Those prescientific people believed in a magical world because of their ignorance and lack of knowledge of nature's phenomenon. They performed rituals and prayed to God for the things they desired. Naturally, whatever they could not explain was attributed to God. On the contrary, it is science, not faith, that discovered the natural phenomenon and solved the mysteries of the past. It is science that has opened our eyes and rewarded the knowledge of nature and life and theories of everything. Along with many others, I would say there is no alternative of science and humanity. Blind faith is just for personal satisfaction, but not the rational way of thinking and pursuing our lives as rational human beings.

Considerable work was done on the necessity of universal religion aimed at achieving the highest level of humanity by David Hockey (2003) in his book *Developing a Universal Religion*. The book covers a wide range of issues such as religion's origin, present-day religions, the purpose of universe and life, moral behavior, science and religion, free will and revelation, and universal religion. He pointed out that a universal religion is one that does not replace existing religions but is, rather, one that might act as an "umbrella" covering the gaps among existing religions and providing moral guidance when none is available or suggesting alternatives when religious differences seem insurmountable. This is somewhat in the line I suggested in the humanity chapter. To get details, I would recommend this book to be read if interested. Here is the website: https://upload.wikimedia. org/wikipedia/commons/5/56/Developing_a_Universal_Religion.

Bringing Scientific Knowledge into Religion

> *When two men of science disagree, they do not invoke the secular arm; they wait for further evidence to decide the issue, because, as men of science, they know that neither is infallible. But when two theologians differ, since there are no criteria to which either can appeal, there is nothing for it but mutual hatred and an open or covert appeal to force.*
>
> —Bertrand Russell

Let me discuss this in simplistic term. Ancient religions used myths to explain the mysteries of the world and nature to primitive people by making and telling stories. But those stories are rejected by modern humans, as those stories are imaginative and not true. As for the modern religions are concerned, the ancient ideas take new shape by the holy books and scriptures that are believed to be the words of God. But the fact is those holy books and scriptures were written by people who had no idea of the real world and the universe but had only primitive understanding of them. They could not grasp that the world circled around the sun and was spinning constantly, and they had no understanding of the solar system that was a part of a much larger galaxy among billions of galaxies. With the knowledge of modern science and technology, we now know that what the writers of holy books saw was completely beyond their understanding. In the absence of adequate knowledge, they had no idea what they were looking at and making conclusions about the god of all things, based on their primitive and incorrect understandings of nature.

The information based on scientific methods has pushed far ahead of ancient misunderstandings and myths. We now know that earth is a tiny dot that hosts all species, lives, including human beings, who, at one time, thought God was the authoritative master of the heavens and the earth, who created all things and who continues to have control over all things, including every human action, every human thought, and every human feeling and aspiration. This is believed to be revealed from God through prophets such as Moses, Abraham, Isaiah, Jesus, Muhammad, and many others. But science has given us new knowledge, understandings, and conclusions about this planet on which we live and the solar system that hosts our world.

Among many others, two elements of science may be looked into to open discussion on the subject in question. The first is that nothing is fixed in the world and in the universe; rather, all things are in motion, and all things are evolving, including human beings. Human beings are not the product of a onetime creative act by an all-powerful god; rather, human beings are always becoming, always arriving. What human beings have become is the result of a long continuous process. Charles Darwin's work on biological evolution pushed forward critical discussion in Christian theology that may be worth noting. The work of Alfred North Whitehead took evolution into the world of philosophy. Charles Hartshorne moved the discussion to theology, who wrote, "Everything, including God, is ceaselessly changing in a dynamic process of creative advance that will never end." This is the good beginning to answer our question.

The second element of science is that there is no beginning and that there will be no end. Soren Kierkegaard, the Danish existential theologian, was cited by Rev. Howard Bess, who noted that according to Kierkegaard, a beginning was not relevant; rather, only the moment was important. As we know, scientists are now explaining and analyzing the outer limits of space in terms of millions of light-years, which are still expanding at accelerating speeds. Thus, one can argue that beginnings and endings of the universe are no longer relevant concepts. However, in the Bible and in the Koran, there are materials and verses about beginnings and endings. The concept of creation of the heavens and earth in seven days relates to end-time theology. But in the light of modern science, this kind of thinking is irrelevant.

There is no clear answer to the question, what kind of religious explanation can relate to science that embraces life that is never static and always in process? The matter of fact is that science continues to progress, but there are still a lot of unknowns that scientists have to search for answers. At the same time, religion at large must have realistic effort and need to create a widely acceptable environment in which people can find the complete and meaningful life, far from unrealistic explanation. Science is an evergreen area of knowledge where the unknown becomes known in a gradual process. One may argue that science is in its infancy. I would rather consider it as never-ending but not complete. On the other hand, religions need to be open, dynamic, creative, and enjoyable, rather than static, compulsive, and fearful to the lives of people. Only then is there a possibility that scientific knowledge can flourish in religions. Rev. Howard Bess hopes that life will be fun, enjoyable, and rich when religious people and scientists are on the same dance floor. Does anyone hope for this to happen? If so, when? These are the open questions.[93]

Reverend A. Powell Davies has explained this subject in a different perspective. According to him, science must enter into the field of religion, or religion must get into the field of science. He further said that religion must be liberal and may be termed as liberal religion and that it must maintain the open-mindedness toward future discovery. It should not be restricted by a creed. In other words, when new knowledge comes, religion must entertain sincerely and take the consequences of it. It suggests that it should accept the advancing truth and accommodate all knowledge and wisdom and always try to know what experience justifies. Only with this kind of liberalism in religion can science meet with religion. Only this kind of religion can keep the door open to scientific advancement without barriers imposed by religious doctrines. This suggests that when scientists go along with religion without abandoning their scientific disciplines, they cannot accept a traditional creed as binding. Scientists want a free field without church's interference and a free and open religion so that they can openly persuade their intellects. It would be rejoicing if traditional denominations will allow this freedom not only to scientists but also to others.

Reverend Powell Davies thinks that scientists do have a need of religion—its basic faith, its moral responsibility, its deeper insights, its wisdom, and its inspiration. Religions need to be genuine and pragmatic. Let us hope that when science and religion meet, their knowledge and resources are mixed together in an honest approach for the common good. What is essential is that religious scholars and preachers truly embrace the scientific methods and the results, and in the same way, scientists take religion to bring into its domain whatever possible through open discussion.

Let us cite some examples of what Reverend Powell Davies has brought forward. Sir James Jeans said from the viewpoint of science, "The universe begins to look more like a great thought than like a great machine." Let us consider this statement comes from liberal religion's perspective, but not as an endorsement of the Apostle's Creed. Sir Arthur Eddington said, "The idea of a universal mind . . . is a fairly plausible inference from the present state of scientific theory." Let us consider that this statement is not in the least the same thing as confirming the doctrine of the Holy Trinity. Prof. Arthur Compton said, "There is something of a nonphysical nature which controls the action of the atom." Let us consider that by this statement, he has not declared his adherence to the *Westminster Confession*. Albert Einstein said that he believes in God, "the god of Spinoza." We should find out what kind of god Spinoza believed in. Religion should not make falsities true and generate fears in our hearts. We must break the bondage of the religious past and become liberated from closed-box thinking to the real, open possibilities of religion. In today's civilized world, religion is required to accept, like science, that the fact of truth is supreme—the free truth in the open world, the truth as experience and open knowledge prove it. Let there be the truth of the heart of people from the knowledge; let there be only truth we seek truthfully that meets our heartfelt needs and desires. To this end, let science and religion meet and mingle. Again, the big question—does anyone hope for this to happen?[94]

Morality

Whenever morality is based on theology, whenever the right is made dependent on divine authority, the most immoral, unjust, infamous things can be justified and established . . . Morality is then surrendered to the groundless arbitrariness of religion.

—Ludwig Feuerbach

It is very common to think that morality depends on the existence of God. For example, some people suggest that there is no right or wrong without God or that atheists who do not believe in God can have no objective basis for their values and that their lives are entirely meaningless. Some people even think that the existence of a moral conscience supports the existence of God.

It is not clear how these claims are to be substantiated. One reason why someone might think that morality depends on God is that he or she accepts, explicitly or implicitly, the divine command theory of morality, which essentially says that "morally right" means "commanded by God" and "morally wrong" means "forbidden by God." However, there is a powerful objection to the divine command theory. This objection derives from a discussion of Socrates in one of Plato's dialogues: (1) is conduct right because the gods command it? Or (2) do the gods command it because it is right?

In the first option, where conduct is right because God commands it, God's commands can be seen as arbitrary, in that God could have given *different* commands just as easily. He could have commanded us to be liars, and then lying, and not truthfulness, would be right. If God says that we should be dishonest, then again this is what we should do. Then goodness of God can be reduced to nonsense, because if we accept the idea that good and bad are defined by reference to God's will, this notion lacks good sense. In the second option, where God commands behavior because the behavior is right, we admit that there is some standard of right and wrong that is *independent* of God's will. What it means is that God seeing or recognizing that a behavior is morally right is quite different from him *making* it morally right. From this, we might ask ourselves, if a behavior is right without the commandment of God, why should God bother to command it? Thus, theological definitions of right and wrong are unnecessary. Adopting a theological definition based on divine command theory would mean accepting the first option but without the goodness of God.

There is another main theory known as the theory of natural law—a system of law that is determined by nature and so is universal. Classically, natural law refers to the use of reason to analyze human nature, both social and personal, and get the binding rules of moral behavior from the natural law. Under this theory, moral judgments are based on reasons. St. Thomas Aquinas, a strong supporter of natural law, emphasized that acting reasonably is not to be contrasted with acting as a theologian. Morality is not a matter of faith but a matter of reason and conscience. Religious considerations do not provide definitive solutions to many of the controversial ethical issues.

One may ask an atheist where he or she derives morality. Well, to answer this question, an atheist can cite science, specifically studies in evolution dealing with reciprocal altruism and selfish gene theory, which show not only *how* but also *why* it is beneficial to be kind to one another. Evolutionary experience we inherit from our ancestors a set of moral intuitions that contributed to survivals in the harsh and changing world. While some of those intuitions still prominently survive, others may be inadequately adapted to the rapidly changing world. So it is our responsibility and foremost task to find out which of them need to be changed. This is why a scientific reevaluation of morality and ethics is necessary.[95]

One might argue that there's a necessary link between religion and morality because of the role religion plays in moral *motivation*. The reason is that if one believes that some god exists and one wants to please that god, then one will behave in a certain way. There is a problem here that one might argue that acting out of fear has little or no moral worth because threats extort but are not capable of imposing moral obligation. One might argue that the belief in a god who will punish and reward us in the afterlife on the basis of our deeds is a necessary component of moral motivation. However, there is a problem here, too, that many atheists and theists behave morally but not out of fear of punishment in the afterlife. One might argue that even if God has given all of us the ability to tell right from wrong, believers have an advantage because of revelation, where God tells the faithful how to conduct their lives. More radically, one could say that morality totally depends on revelation. But there are problems that we have to consider as to what counts as sacred texts and what their teachings are. Which texts or oral traditions constitute God's message to us? For example, is it the Hebrew Bible? Or is it the Christian Bible? If so, which version? Or is it the Quran? Or is it any other religious sculptures? Believers justify their action by appealing to their respective sacred books. Now let us propose something. If religions per se tell us that killing is immoral, why don't they remove the relevant passages from their sacred book? Well, they don't do it because not one word or passage of the holy books can be changed, because it is the word of God. Now going back to the fundamental that we have to believe what believers decide to interpret their sacred books, which gives injunctions that we should be killed if we do not follow the commands interpreted by them. This is not accepted by the nonbelievers and open thinkers.

We live in a world that wants the fruits of scientific labor but refuses the mental discipline of scientific rationality. For all religions, there is an "us" and a "them." In this dichotomy, what is important to notice is that all the rituals and the associated religious practices are just the means of encompassing and justifying inclusiveness within group boundaries, including practiced hostility. It is no accident that apart from some good ethical philosophy, the history of world religions is a history of violence, hatred, and intolerance.

Faith, by definition, is not rational but a belief in the absence of investigation that requires logical explanation and proof. If every assertion were subject to question, the faithful and believers would have to admit that they hold their beliefs without rational basis. If the free contest of open ideas requires promoting in the public domain, religious belief would be subject to the scrutiny of scientific rationality. In every religious tradition, there is orthodoxy with priests, ministers, rabbis, mullahs, and saints to enforce it, and strong effort is made to suppress dissent. The irony is that we still live at a time when more and more people are demanding that unpopular ideas must be suppressed. Speaking freely and openly of rationality, logically meaningful, and of progressive ideas is now an invitation to serious trouble that affects personal and social life.

Faith, like superstition, prevents moral expression and action. Those who fail to understand how the world works with rational sequences are incapable of adopting informed choices, as they see only the demons and angels, miracles, curses, and religious blindness. They are unable to take responsibility for their actions because they lack intellectual and emotional maturity. This is why Friedrich Nietzsche hated religion so much. Nietzsche despised weakness. He did not think of weakness primarily in terms of physical strength. Rather, he was referring to quality of character. We must be mentally weak to be controlled by religion and its orthodoxy. In fact, religion is one of the most effective institutions of social control; believe it or not. Although organized religions sometimes are not in agreement with governments, the faithful are merely foot soldiers of the spectacle.

It is understood that foot soldiers' job is to instigate widespread fear among people. They see sin everywhere, and sin brings punishment. To instigate fear, they unveil an angry god to be an effective rhetorical tool. The question is, why fear? Because fear prevents us from being open to the varieties of beauty and pleasure around us. When we are afraid, we seek comfort in warm, enclosed spaces—literally and figuratively. The fearmongers hope to send us quickly toward the safety of their prison. The strategy works too. We can now understand why so many people turn to religion as a refuge from complexity. Both function, albeit by different rituals and with different ideologies, to create the illusion of security. Believers are protected from the ugly truths of the real world. The cost, of course, is the opportunity to explore what the world has to offer. Children must do what they are told; so must the faithful. At least children know that the power exercised over them keeps them from enjoying themselves. Those who surrender responsibility for their own moral action lack that insight. To them, slavery is freedom and religious faith is nothing but liberating self. This is utter nonsense. Rationalists and open thinkers see this example of mental gymnastics as self-delusion.

In contrast to the believer or faith, people can be artists who are exemplars of courage and creativity. Creativity requires a boldness and courage, which an artist possesses, that can be strongly exercised to everyday living. An artist must have a scientific rationality because of his/her sense of imagination and discovery. Otherwise, his/her work will be dull and liveliness and have lack of flavor. The artists can be distinguished by two different meanings. The first refers to a kind of person who produces art—that is, paints a painting, writes a poem, composes a musical piece, and so forth. This is the artist as a technician, someone who is skilled in technique or craft of artistic design. Here we refer to the artist not in this type but of the second type—an aesthete. The artist is an aesthete. When technique is a set of skills that may be acquired through practice, aesthetic awareness needs to be cultivated by a difficult discipline. It requires a certain mind-set that is quite different from ordinary awareness. It is the sensitivity to the subtleties of beauty and sensual pleasure. It is the familiarity with the positive and negative aspects of stimulation and feeling and, thus, appreciation of both forms. Whereas the artist as craftsman might produce a religious object of devotion, the artist as aesthete is diametrically opposed to the believer.

The artist as aesthete is primarily open to the wide range of experiences of humanity. All religious traditions are basically antipleasure, which is why religious law is so obsessed with sexuality. Thinking rationally, you can't be fully human if you are unrelentingly hostile to pleasure and devoid of the real beauty and flavor of life. The aesthete is defined by openness to the sensual world just as the believer is defined by closeness to it as they live in the religious box. Lacking the proper appreciation for humanity, believers are subject to be motivated to committing unspeakable acts. The closeness to pleasure and not adopting new ideas is a necessary condition for the kind of suffering that makes a young person susceptible to irrational persuasion and acts. This core mind-set keeps them away from beauty and pleasure and turned toward engaging into hatred and violence.

More than 80 percent of the world population is adhered to some form of religious belief. But it is curious to know how many believers are ignorant of the history of their respective religions. Majority of them know only the stories told in their scriptures as genuine and authentic. They never search for the truth that lies beyond stories of scriptures. As they live in the religious box, they are devoid of the open world concept completely. But humanity is more than religion, and at the times of history, humanity would be better off without religion. Everyone who appreciates the good, the true, and the beautiful gifts of nature has a duty to challenge the religious superstition in every way. It is not enough to be irreligious; we must use our knowledge, intelligence, and even critique to expose religion for what it stands for. This is a duty that we should take very seriously. To speak rationally, it is increasingly dangerous in a society that is defined by fear. Somewhere, there are fanatics who think killing people who oppose religious views will be their ticket to heaven. We must stop those fanatics. It won't be easy, because shifting the momentum of history never is. In a sense, ours is a fool's errand, but it is not folly. We may not actually hasten the demise of religion (that would be too much to hope for), but we can slow down further growth and eventually slide to the bottom. As we stand today, there are only limited or zero spaces of freedom by living in religious box. This is what we fight for—coming out of the box. Our work exposing the contradiction between religion and morality and humanity at large will hopefully preserve our freedom to think openly.[96]

We know that in the earliest Western legal systems, the existence of human rights is derived from secular logic, rationality, and humanitarianism. It is found in the book of Hugo Grotius, *De Jure Belli ac Pacis*, in the seventeenth century. The book became famous for codifying mortality without any need of laws and divinity, based on reason and humanitarianism. Since then, human rights have become an increasingly powerful tool used in the fight against arbitrary oppression, intolerance, and unjust mob rule. We can also refer Jack Donnelly, a political theorist who specializes in human rights and is the author of *Universal Human Rights in Theory and Practice*. He emphasized that the source of human rights is man's moral nature, no less, and that internationally recognized human rights do not depend on any particular religious or philosophical doctrine. It is to be noted that truths are proclaimed by human rights documents themselves—"such as those in the covenants that *these rights derive from the inherent dignity of the human person* or in the Vienna Declaration that *all human rights derive from the dignity and worth inherent in the human person.*" In the context of nonsectarian and postcultural view, the human rights come from moral understanding and thinking, but not from any particular religious or ethical philosophy. Given the fact that such thinking grows inside particular cultures, those cultural biases have been eliminated from the world's human rights documents and introduced a purely secular and universal concept of human rights.

Evidence suggests that the secular human rights approach is correct. Social and moral development is higher in countries that are less religious. As religiosity increases, a country suffers from conflicts with human rights, more problems with tolerance of minorities, and religious freedom and problems with gender. It has also been observed that the countries where population are less religious or not religious are socially, culturally, and economically advanced, morally good, and less corrupt than those countries where populations are deeply religious. Prof. Victor J. Stenger pointed out the fact that religious codes cannot be the origin of people's moral thoughts. The moral thoughts are remarkably similar in people with different religious concepts or without any such concepts. Even religious people's thoughts about morals are constrained by intuitions they share with other human beings more than official codes and models. Religious nobles and preachers tell us that any universal moral standards can only come from one source—their particular god. Otherwise, standards would be relative, depending on culture and differing across cultures and individuals. The data suggests that the majority of human beings from all cultures and religions or no religion agree on a common set of moral standards.

Thus, universal moral norms do exist.[97] Our culture, individual psychology, careful and harmless actions, and secular thinking are the true sources of morality. As Pascal Boyer said, "Our evolution as a species of cooperators is sufficient to explain the actual psychology of moral reasoning and requires no special concept of religious agent, no special code, no models to follow, even though you can easily insert them in moral reasoning that would be there in any case."[98]

"A man's ethical behavior should be based effectually on sympathy, education, and social ties and needs. No religious basis is necessary. Man would indeed be in a poor way if he had to be restrained by fear of punishment and hopes of reward after death," said Albert Einstein ("Religion and Science," *New York Times*, 1930). There have been people who have had a morality but no religious beliefs. Bernard Williams, an English philosopher, stated that the secular utilitarian outlook* is "nontranscendental and makes no appeal outside human life, in particular not to religious considerations." He argued, "Either one's motives for following the moral word of God are moral motives, or they are not. If they are, then one is already equipped with moral motivations, and the introduction of God adds nothing extra. But if they are not moral motives, then they will be motives of such a kind that they cannot appropriately motivate morality at all . . . We reach the conclusion that any appeal to God in this connection either adds to nothing at all or adds the wrong sort of thing."

In "Euthyphro dilemma," Socrates refuted the idea that requires religion. Similar reason was described by Peter Singer as saying, "Some theists say that ethics cannot do without religion because the very meaning of 'good' is nothing other than 'what God approves.' Plato refuted a similar claim more than two thousand years ago by arguing that if the gods approve of some actions, it must be because those actions are good, in which case it cannot be the gods' approval that makes them good. The alternative view makes divine approval entirely arbitrary: if the gods had happened to approve of torture and disapprove of helping our neighbors, torture would have been good and helping our neighbors bad. Some modern theists have attempted to extricate themselves from this type of dilemma by maintaining that God is good and so could not possibly approve of torture,

---

* A popular ethical position wherein the morally right action is defined as that action that effects the greatest amount of happiness or pleasure for the greatest number of people.

but these theists are caught in a trap of their own making, for what can they possibly mean by the assertion that God is good?"

A Harvard University humanist chaplain, Greg Epstein, argued about the question of whether God is needed to be good and said that this question does not need to be answered and needs to be rejected outright. He also said that one *can't* be good without belief in God, which is not just an opinion; it is a prejudice. It may even be discrimination. This argument is within the purview of the *Westminster Dictionary of Christian Ethics*, which states that religion and morality are to be defined differently and have no definitional connections with each other. Conceptually, morality and a religious value system are two distinct kinds of value systems or action guides. Others share this view. Singer states that morality is not something intelligible only in the context of religion. Atheistic philosopher Julian Baggini strongly emphasized the fact that there is nothing to stop atheists believing in morality, a meaning for life, or human goodness. It is as capable of a positive view of other aspects of life as any other belief. He also said that morality is more than possible without God; it is entirely independent of God. In other words, atheists are not only more than capable of leading moral lives but also able to lead more moral lives than religious believers who confuse divine law and punishment with right and wrong.

A very popular atheist and writer, Christopher Hitchens, remarked on the program called *Uncommon Knowledge* (a TV show) as saying, "I think our knowledge of right and wrong is innate in us. Religion gets its morality from humans. We know that we can't get along if we permit perjury, theft, murder, rape. All societies at all times, well before the advent of monarchies, certainly have forbidden it." Philosopher Daniel Dennett said that the idea that people need God to be morally good is extremely harmful, yet it is a popular myth. According to him, it is a falsehood that persists because churches are much better at organizing people to do morally good work. In his words, "What is particularly pernicious about it [the myth] is that it exploits a wonderful human trait. People want to be good. They want to lead good lives . . . So then along come religions that say 'Well, you can't be good without God' to convince people that they have to do this. That may be the main motivation for people to take religions seriously—to try to take religions seriously, to try and establish an allegiance to the church—because they want to lead good lives."

There are evidential findings that suggest that morality does not depend on religion but that, rather, it is the instinct in the brain that operates morality, even in the cases of animals' nature that shows moral behavior. An example is the studies of the complex systems of altruism and cooperation that operates among social insects and the act of altruistic sentinels by some species of birds and mammals that risk their own lives to warn and save the rest of the group from imminent danger. Sociologists have recently found that some of the world's most secular countries, such as those in Scandinavia, are among the least violent, best educated, and most likely to care for the poor. Scientists are beginning to document the fact that, though religion may have benefits for the brain, secularism and humanism also may have the same benefits.

A study conducted in 2012, which tested prosocial sentiments and was published in the *Social Psychological and Personality Science* journal, found that nonreligious people had higher scores, showing that they were more inclined to show generosity in random acts of kindness, such as lending their possessions and offering a seat on a crowded bus or train. Religious people also had lower scores when it came to seeing how much compassion motivated participants to be charitable in other ways, such as in giving money or food to a homeless person and to nonbelievers. Other studies conducted in various countries found no relationship between faith and crime; rather, the evidence surrounding the effects of religion on crime is found to be varied and inconclusive. Phil Zuckerman, in his book *Society without God*, noted that Denmark and Sweden, which are the least religious countries in the world, enjoy among the lowest violent crime rates and the lowest levels of corruption in the world. Many other studies have been conducted on the subject as well.

A study by Gregory S. Paul in 2005, which was published in the *Journal of Religion and Society*, found that higher rates of belief in and worship of a creator correlate with higher rates of homicide, juvenile and early adult mortality, STD infection rates, teen pregnancy, and abortion in the prosperous democracies. The study also found that in all secular developing democracies, a century's long-term trend has shown that homicide rates dropped to a historical low, with the exception of the United States (which has a high religiosity) and theist Portugal. Further investigations were done by Gary Jensen on Paul's study and found that a complex relationship exists between religiosity and homicide, with some dimensions of religiosity that encouraged homicide, while other dimensions discouraged it.[99]

Chapter Two Conclusion

Freethinkers consider themselves as people who are the actors on their lives on earth, and, thus, they are responsible for the good and bad things that happen on earth; that is, the natural world and humanity's place in that world are the focus of their intellect and convictions. Freethinkers do not look for the supernatural or to a deity to solve their problems. When someone does not associate with any religion or religious doctrine, they can be called by different terms such as atheist, agnostic, pantheist, freethinker, rationalist, skeptic, naturalist, secularist, etc. The question is, why do people (being born in one the religions family tradition) leave or disassociate with the religion in which they were or wholly disassociate with any other religions? Here are some explanations. For intelligent freethinking people who are not suffering from any major life-concerned issues and who do not have low self-esteem, religion is not meaningful in life. Then think that the decision to formally associate with a religion is just burdening the mind with a heavy load of false notions.

Religion blocks people's independent thinking. We are taught by the religion on what to believe instead of encouraging us to think freely beyond the holy book. This does not help us for spiritual growth. Can we not call religion the off switch of our mind? People's intellect is a better instrument of not only spiritual growth of mind but also every aspect of human endeavor than any religious teachings.[100] Steve Pavlina thinks that religious teachings are invariably mysterious, confusing, and internally incongruent. By injecting this kind of confusing and internally conflicting information, our logical mind (i.e., neocortex) is overwhelmed. As a result, our logical mind is cut off from logic because it can't find a pattern of core truth beneath all the illogical contents. So without the help of the neocortex, we devolve to a more primitive (i.e., limbic) mode of thinking.[101] I am of the same opinion of what Pavlina has explained. I have discussed similar neurological aspects of religious beliefs in chapter 1.

A recent review of sixty-three studies showed that there is a moderate negative relationship between intelligence and religiosity (Zuckerman, Silberman, and Hall 2013). Openness to experience, along with intelligence, is also associated with greater *general knowledge* of the world. This may be because people who are high in openness to experience are intellectually curious and, therefore, motivated to learn new things about the world.[102] Evidence suggests that people who rely upon their intellect to solve problems have a greater chance of living satisfying lives than those who depend on the supernatural hope that an unseen sky god will somehow save them from their troubles. Research also suggests that life satisfaction is three times more likely to correlate with a belief in science and technological progress than belief in religious doctrine. The findings come from the World Values Survey (WVS) from seventy-two countries conducted in 2014.[103] The findings shadow light on the tenet that a belief in scientific-technology progress is a key feature of optimism for future humanity. This in fact turns out that people who believe in the progress of science and technology tend to be happier than the religiously faithful people. The world's happiest countries are also the world's least religious countries, according to a new report released by the World Happiness Index.[104] In other words, we may say that humanists are happier than those who have strong affiliation with religion. God likes people who are humanists because they promote love, not hate. Let me quote this:

> God didn't draw up religious barriers to separate us from each other. Man did. And on top of that, no father would like to see his children fighting or killing each other. The Creator favours the man who spreads loves over the man who spreads hate. A religious title does not make anyone more superior over another. If a kind man stands by his conscience and exhibits truth in his words and actions, he will stand by God regardless of his faith (Suzy Kassem).

Believing in God does not always mean believing in religion. While some would argue that only one religion leads to God, others might say that all religions lead to God. But there are other groups who would argue that no religion leads to God, because the religions confine within a box along with their dogmas, which suggests that religions fail to convince everyone about the true reality of God. I think that God is much bigger than any religion. Religion is like a corporate identity that ends up speaking, sounding, practicing, and believing what corporate branded identity has established. When someone steps out of line and has questions that don't fit that corporate identity, there are chances that they will be silenced, cast out, or even killed.

Religious scriptures do contain some commonsense moral education; it's true. But religions are also demonstrating with celebrations of cruelty, slavery, and violence, which suggest that they are not the best guides on how to act morally or ethically. Most religious people believe that morality depends on religion. It is not true. In reality, though, morality is a human phenomenon. It is the product of our evolution as a social species, and good and evil have nothing to do with any god or religion. People who are not strongly religious (by not blindly following religious dogmas and scriptures) make responsible moral choices by their own conscious, reasons, and rational judgment and know the consequences of what they do. In many situations, religious people make personal choices about moral practices they need to adopt even when those are not necessarily guided by the religion.

We are a naturally kind and moral species because of millions of years of maintaining group harmony for survival. Morality is not exclusive to religion. Everyone has their own moral standards, including nonreligious people—atheists, agnostics, and freethinkers. There are arguments that religion-free people are more moral than the religious people. Global population demographics suggest that atheism and agnosticism are far more common among scientists, philosophers, and distinguished thinkers and that the most peaceful and equitable societies on earth are least religious. Evidence also suggests that morality is higher in the general population of the secular countries where people are less religious or nonreligious compared to those with religious population majority countries. Moreover, secular and less religious countries are relatively much better in terms of economic well-being, social justice, equality, happiness index, and other socioeconomic parameters compared to religious majority countries in the world.

As citizens of the global society, what we experience is that many people think they like to neither take the identity of religion nor retain or seek its membership. Rather, they are happier to have a relationship that is unique with God in the way they want and live in the real life. They like to live, think, walk, and speak freely every day with an open mind to learn new things, to gather new knowledge, and, with an open heart, to connect authentically and rationally with the world around them.

CHAPTER THREE

Humanism

Life has no meaning a priori . . . It is up to you to give it a meaning, and value is nothing but the meaning that you choose.
—Jean-Paul Sartre

In dark ages people were best guided by religion, as in a pitch-black night a blind man is the best guide; he knows the roads and paths better than a man who can see. When daylight comes, however, it is foolish to use blind old man as guides.
—Heinrich Heine

History, Knowledge, and Values

Humanism Defined

Humanism is described as a philosophy, a worldview or life stance that focuses mainly on nature and humans and rejects the supernatural. Humanism is positive, ethical, and democratic. It is underpinned by a commitment to rational inquiry and the scientific method. Humanists believe and accept the fact that this is the only life we can know of and that it is our only time to live and enjoy to the fullest. It is our individual and collective responsibility to cooperate and work together in trying to find solutions to the world's problems and to preserve the planet now and in the future. The word *humanism* is derived from the Latin concept *humanitas* in the nineteenth century. Historians agree that the concept predates the label invented to describe it, encompassing the various meanings ascribed to *humanitas*, which included both benevolence toward one's fellow humans and the values imparted by *bonae litterae*, or humane learning (literally "good letters"). *Humanitas* meant the development of human virtue, in all its forms, to its fullest extent. The term thus implied the modern word *humanity* is associated with not only such qualities as understanding, benevolence, compassion, mercy but also such more aggressive characteristics as fortitude, judgment, prudence, eloquence, and even love of honor.

In short, humanism called for the comprehensive reform of culture, the transfiguration of what humanists termed the passive and ignorant society of the Dark Ages into a new order that would reflect and encourage the grandest human potentialities. All the Renaissance writers who cultivated *humanitas* and all their direct descendants may be correctly called humanists.

While slow in growth in the making of permanence itself, humanism, in the large measure, established the climate and provided the medium for the rise of modern thought. The major developments in literature, philosophy, art, religion, social science, and natural science had their basis in humanism. Notable people in all fields regularly made use of humanistic eloquence to further their causes. Prof. Robert Grudin (University of Oregon, Eugene, author of *The Grace of Great Things* and others) said that the modern awareness (that is, the sense of alienation and freedom applied both to the individual and to the race) derives ultimately from humanistic sources. Apart from its skepticism and inner conflicts, the humanistic movement was heroic and remarkable in its aspirations. Humanity's moral values and programs formed the basis for lives that are remembered with admiration, and we should follow and practice not only at the individual level but also in the community and state levels. The core of humanism has probably been best expressed by the philosopher Baruch Spinoza when he wrote, "Peace is not an absence of war. It is a virtue, a state of mind, a disposition for benevolence, confidence, and justice." This vision is a universal compass that we must tailor to the realities at all times.

Brief History of Humanism

A Latin grammarian, Aulus Gellius (c. 125–c. 180), said that in his day, *humanitas* was commonly used as a synonym for philanthropy— or kindness and benevolence toward one's fellow human being. Gellius maintained that this common usage was wrong and that model writers of Latin, such as Cicero and others, used the word only to mean what we might call humane or polite learning, or the Greek equivalent *paideia*. Gellius became a favorite author in the Italian Renaissance, and in fifteenth-century Italy, teachers and scholars of philosophy, poetry, and rhetoric were called and called themselves humanists. Modern scholars, however, point out that Cicero (106–43 BC), who was most responsible for defining and popularizing the term *humanitas*, in fact, frequently used the word in both senses, as did his near contemporaries. For Cicero, a lawyer, what most distinguished humans from brutes was speech, which, allied to reason, could (and should) enable them to settle disputes and live together in concord and harmony under the rule of law. Thus, *humanitas* included two meanings from the outset, and these continue in the modern derivative, *humanism*, which, even today, can refer both to humanitarian benevolence and to scholarship.

During the French Revolution, and soon after, in Germany, the so-called Left Hegelians, Arnold Ruge and Karl Marx (who were critical of the close involvement of the church in the repressive German government), *humanism* began to refer to an ethical philosophy centered on humankind as opposed to institutionalized region. The designation religious humanism refers to organized groups that sprang up during the late nineteenth and early twentieth century. Although it is similar to Protestantism, it is centered on human needs, interests, and abilities rather than the supernatural.

In pre-Socratic time (sixth century BCE), Greek philosophers Thales of Miletus and Xenophanes of Colophon were the first in the region who attempted to explain the world in terms of human reason rather than myth and tradition. These Ionian Greeks were the first thinkers who openly said that nature is available to be studied separately from the supernatural realm. Other influential pre-Socratic rational philosophers include Protagoras, known for his famous dictum "Man is the measure of all things," and Democritus, who proposed that matter was composed of atoms. Little of the written work of these early philosophers survived, and they are known mainly from fragments and quotations of other writers, principally Plato and Aristotle. The historian Thucydides (who was renowned for his scientific and rational approach to history) was much admired by later humanists. In the third century BCE, Epicurus was remembered for his notable expression of the problem of evil, lack of belief in the afterlife, and human-centered approaches to achieving *eudaimonia* (human flourishing). He was also the first Greek philosopher to admit women to his school as a rule.

Humanism as a philosophy existed in Asia. Approximately 1500 BCE, human-centered philosophy that rejected the supernatural can be found in the Lokayata system of Indian philosophy. The Nasadiya Sukta, a passage in the Rig Veda, contains one of the first recorded assertions of agnosticism. In the sixth century BCE, Gautama Buddha expressed, in the Pali literature, a skeptical attitude toward the supernatural. Instance of ancient humanism as an organized system of thought is found in the Gathas of Zarathustra, composed between 1000 BCE and 600 BCE in Iran. In China, Huangdi (the Yellow Emperor) is regarded as the humanistic primogenitor. Sage kings such as Yao and Shun were the humanistic figures. The famous saying of King Wu of Zhou has been recorded as "Humanity is the Ling [efficacious essence] of the world [among all]." Duke of Zhou (a member of the Zhou Dynasty who played a major role in consolidating the kingdom established by his elder brother King Wu) was regarded and respected as a founder of Rujia (Confucianism) and is especially prominent and pioneering in humanistic thought. The Silver Rule of Confucianism is an example of ethical philosophy based on human values rather than the supernatural. In the Taoist and Confucian secularism, elements of moral thought devoid of religious authority show some resemblance to the modern concept of secularism.

Many medieval Muslim thinkers and philosophers pursued humanistic, rational, and scientific discussions and debates in their search for knowledge, meaning, and values. Many writings on love, poetry, history, and philosophical theology by Islamic writers show that medieval Islamic thought was open to the humanistic ideas of individualism, secularism, skepticism, and liberalism. Those were the emphasis on freedom of speech, which was perhaps one of the reasons the Islamic world flourished during the Middle Ages.[105]

It is also suggested that humanism as a philosophical and literary movement originated in Italy in the second half of the fourteenth century and spread all over Europe. As an atheistic theory, it was conceived in the seventeenth century, but as a theistic pragmatic theory, it was introduced indirectly around 200 BC at the time of *Vedas* and *Upanishads* in India. The prayer "Sarvatra Sukhinah Santu Sarve Santu Niramayah," meaning "Let all be happy here and let all enjoy full health"—the Vedic sages echoed this universal welfare. The earthly life constitutes the central concern for the Vedic Aryans. The sacrificial fire rites, which were evolved during the Vedic period, had social welfare as its motto; the motive was to prepare the land for agriculture for abundance and welfare of the human race.

The latter half of the nineteenth century witnessed Hindu Renaissance pioneered by Brahm Samaê of Raja Ram Mohan Roy and Arya Samaê of Dayanand Saraswati, finally blossoming into Vedantic Hinduism of Vivekananda. Vedantic Hinduism stresses the importance of service to the weak and the needy as its practical aspect. The salient theme is "Society is the greatest where the highest truths become practical." Humanism has undergone significant development at various levels and forms in the West as well as in the East. But there is a basic difference between Western humanism and Eastern humanism. While the former is atheistic in content because of the conception of God as the creator, the latter is the Vedantic humanism, which is not atheistic.[106]

Renaissance and Reformation of Humanism

Humanism is an attitude of thought that gives primary importance to human beings, and its outstanding historical example was the period of Renaissance humanism from the fourteenth to sixteenth centuries, rediscovered and developed by European scholars of classical Latin and Greek texts. During that time, much of the wisdom of the ancient world was lost or destroyed, in which intellectual life was dominated by religion and theology. It is often called the Dark Ages for this reason. In opposition to the religious authoritarianism of medieval Catholicism, strong emphasis was placed on human dignity, beauty, potential, and every aspect of culture in Europe, including philosophy, music, and the arts. As a result of this humanist emphasis on the value and importance of the individual, it influenced the reformation and, in turn, brought about social and political change in Europe. In fact, there was little or no freedom for most people, and religious dissent, or heresy, was harshly punished. However, from the ninth century onward, important European cultural people laid the foundation for the Renaissance.

The Renaissance was an important period of intellectual and artistic development. Intellectuals and thinkers were seeking ideas from intellectuals all over the world. Many Arab scholars in the areas of mathematics, astronomy, and medicine were brought to Europe. It was an era of exploration and discovery. However, the church was often hostile to new scientific ideas, which it saw as threatening. The Polish astronomer Copernicus (1473–1543) suggested that everything in our solar system revolved around the sun and discarded the traditional idea that everything revolved around the Earth. This was opposed by the church, as many others, such as Giordano Bruno and Galileo, who publicly accepted or developed Copernicus's ideas. In England, Francis Bacon (1561–1626) developed a theory of scientific method and recommended thorough collection of data before drawing conclusions.

The visual arts were characterized by a growing realism and were accepted by people. The use of perspective and drawing from nature were reflected in the arts, and arts gradually became more diverse rather than only religious in the arts of key subject matter. Famous Italian artists in that period include Uccello, who was an early user of perspective in his paintings, and Leonardo da Vinci, who was known as Renaissance man for his wide range of interest and knowledge of art. Leon Battista Alberti (1404–72) was another Renaissance man for his wide-ranging abilities and interest regarding nonreligious ethics, a view of citizenship, and the architecture of cities that were very secular and modern. His idea was that the city must provide the best possible setting for its citizens and that the architect of the city must serve the needs of man with dignity.

Playwrights such as Christopher Marlowe and William Shakespeare developed a new kind of theater that was more secular, more realistic, and more interesting in human psychology and emotions. The invention of the printing press at that time is worth noting, as it was an effective means of dissemination of ideas more easily. Aphra Behn, the first Englishwoman to earn her living by writing, wrote critically about religion and slavery.

While the process and ideas began during the Renaissance in the seventeenth century with a growth in religious dissent, the eighteenth century was a period of intellectual discovery, and the dissent (religious, political, and social) became more open. A few enlightened rulers, such as Frederick the Great of Prussia, were patrons of radical writers and thinkers, fostering the growth of new ideas. Notable figures like the radical philosopher and campaigner Thomas Paine influenced the French and American revolutions that took place at the end of the century, and Mary Wollstonecraft pioneered feminist ideas in her writings. Atheism was uncommon and persecuted, but criticism of organized religion and traditional religious beliefs was widespread, often coupled with radical political ideas. Religious skepticism became more common in the eighteenth century as a consequence of the development of a more scientific view of the universe. The Scottish philosopher David Hume wrote very critically about miracles and religion. In France, a group of radical and freethinking philosophers, who were highly influential, expressed their liberal, materialist, empiricist, and naturalist ideas and their skeptical attitude to religion. Their ideas influenced the course of the French Revolution. In Germany, the philosopher Immanuel Kant revolutionized the studies of metaphysics and ethics.[107]

The nineteenth century is often thought of as being a pious age. On the other hand, it was a period of skepticism and renunciation of faith for many thinkers as well. The intellectual and religious atmosphere was already changing by the beginning of the nineteenth century. Humanist thinking developed rapidly in the nineteenth century because it was closely associated with new scientific thinking and discoveries. Darwin's ideas and new biblical research and scholarship coming from Germany provoked a crisis of faith in many Victorian intellectuals. Darwin's defender T. H. Huxley coined the word *agnostic* to describe his belief that there were things that we could not possibly know. The most notable publication of the nineteenth century was Charles Darwin's *Origin of Species*. Published in 1859, it explained the origin of man by describing evolution by natural selection over millions of years and confirmed what many had suspected. Upon learning how life on earth evolved and realizing that there was no need for a creator, many people became agnostics, though others continued to prefer the biblical account. There were many nonreligious scientists who were motivated by the desire to gain more knowledge about the workings of the universe around them and to help improve the condition mankind was in.[108]

Humanism, as it was conceived in the early twentieth century, rejected the revealed knowledge, religion-based morality, and the supernatural. Thus, there was a decline in religious belief and an increase in secularization. Fewer people in Europe are actively religious, and people are free to declare their disbelief in gods with little fear of reprisal or social disadvantage. Most twentieth-century philosophers, such as Sir Karl Popper, A. J. Ayer, G. E. Moore, Mary Warnock, Jean-Paul Sartre, Simone de Beauvoir, Wallace Matson, Antony Flew, and Peter Singer, have worked on the assumption that morality is independent of religious faith. Because of their belief that this world is the only one we have and that human problems can only be solved by humans, humanists have often been very active social reformers. There have been huge developments in science and medicine, which have affected people's lives and the way they think. As more and more people around the word acquire education, understanding of science has become much more widespread, and once-controversial ideas such as Darwin's theories about evolution are generally accepted.

Thanks to the relatively new sciences of sociology, anthropology, and psychology, for the fact that our understanding of human nature and society has developed rapidly. Many scientists were and are humanists. Some such as Sir Arthur Keith (1866–1995), Scottish scientist and anthropologist J. B. S. Haldane, Sigmund Freud, Sir Julian Huxley, and John Maynard Smith did much in the twentieth century to spread understanding of science, of human nature, and of evolution. Albert Einstein, who worked out the theory of relativity and one of the greatest achievements of the human intellect, was essentially a humanist. Scientific and medical progress has produced new ethical dilemmas, and traditional religious teachings have not always been able to rise to the challenge.[109] It is extremely important for us to believe in the dynamic nature of humanism that takes curiosity, creativity, learning, and pursuing knowledge at the forefront of the human experience. We see technology as a remarkable work of human endeavor that makes our life easier, more meaningful, efficient, conscience, and full of variety. We live in a world where we can enrich our lives with variety of our own creation. Such a world can only exist when people are free, where freedom of speech, freedom of belief, and freedom of choosing life are open to all.

Humanist Manifesto

There are three *Humanist Manifesto*s worldwide. The first *Humanist Manifesto* was issued by a conference held at the University of Chicago in 1933. Signatories included the philosopher John Dewey, but the majority were ministers (chiefly Unitarian) and theologians. They identified humanism as an ideology that espouses reason, ethics, and social and economic justice, and they called for science to replace dogma and the supernatural as the basis of morality and decision-making.[110] The second one is named as the *Humanist Manifesto II* (1973), and the third one is named as *Humanist Manifesto III*, also known as *Humanism and Its Aspirations* (2003). For discussion, one can visit https://en.wikipedia.org/wiki/Humanist_Manifesto.

International Humanist and Ethical Union (IHEU) is the world union of 117 humanist, rationalist, irreligious, atheistic, bright, secular, ethical culture and free thought organizations in thirty-eight countries. The Happy Human is the official symbol of the IHEU, as well as being regarded as a universally recognized symbol for secular humanism.

The following is according to the IHEU's bylaw 5.1:

> *Humanism is a democratic and ethical life stance, which affirms that human beings have the right and responsibility to give meaning and shape to their own lives. It stands for the building of a more humane society through an ethic based on human and other natural values in the spirit of reason and free inquiry through human capabilities. It is not theistic, and it does not accept supernatural views of reality.*

Humanism is about maximizing the safety, well-being, and potential prosperity of all people in our society, putting emphasis on human values, on human rights, and on humane behavior toward one another. It is the notion that only through reason, respect, empathy, and compassion can we create a more beneficial society for all with justice and equality. It is a set of principles that can appeal to the whole of society regardless of their personally held religious beliefs or atheism, as it is about designing society through optimizing human collaboration around what benefits us all in this life, here and now, in the shared social space. In today's climate, financial meltdown, global pollution, corruption of the democratic process, and inequality across societies, we need, more than ever, a plan for our collective future, for our very survival, for our continued well-being, and for the opportunity to prosper in all segments of our society. These democratic values and principles should be the political philosophy, economic model, and vision for a better world; this is democratic humanism.[111]

Humanists, in approaching life from a human perspective, start with human ways of comprehending the world and the goal of meeting human needs. These lead to tentative conclusions about the world and about relevant social policies. Because human knowledge must be amended from time to time and because situations constantly change, human choices must change as well. This renders the current positions on social policy the most adaptable part of the humanist philosophy. As a result, most humanists find it easier to agree on basic principles than on tentative conclusions about the world and easier to agree on both than on social policies. Clarity regarding this point will erase many prevalent misunderstandings about humanism.

While not all religions will accept each of the humanist principles, there are many ideas upon which there can be common agreement. Humanists believe that there are many problems such as pollution, war, starvation, climate change, and continued improvements in food distribution. These are examples of human-made problems that can be solved through human efforts, regardless of our individual religious affiliation. We got to think seriously that the responsibility for our lives and the kind of world in which we want to live in is ours only. We should adopt this vision and have firm commitment to act, and we know that humanity is the supreme cause of everything we do and live. And last but not the least, humanity has the power and ability to advance to its highest degree.

It may be helpful to remind ourselves that the historical humanistic ideas of compassion and caring that have pervaded religions throughout the centuries have their origins in early Greek philosophy. These same humanistic principles have helped keep religion relevant, personal, and dynamic. A religion devoid of humanistic principles is not likely to flourish in the Western world. The philosophy of humanism extends early humanistic principles into modern life in the twenty-first century. An extremist religious movement such as ISIS (Islamic State of Iraq and Syria), now also known as the Islamic State, which aims at forming a global caliphate, is the one that Western civilizations fear most because they are lacking the humanistic values of caring, sharing, and respecting our fellow humans. Religions have been improved and can be improved with the acceptance of more humanistic values and ethics.[112]

Knowledge, Ethics, and Value of Humanism

Humanists are those who recognize that they are the most curious and capable curators of knowledge on earth. To gain knowledge, humanists use reasons and experience to understand the world. And they may create the great artistic fruits of humankind to enhance their emotional palettes, deepen empathy, and enrich understanding. However, humanists reject any reliance on blindly received authority or on dogma or on any divine revelation. Humanists are the ones who recognize that they are, by far, the most sophisticated moral actors on earth. They can grasp ethics. They are not being the only moral subjects (other animals deserve moral consideration too). But human beings have a unique capacity for moral choice, such as acting in the interests of welfare, advancement, and fulfillment, or against it. To act right, they must take responsibility for themselves and others, not for the sake of preferential treatment in any afterlife (even if we believed in it, that motivation wouldn't make our actions good!) but because the best they can do is to live this life as brilliantly as they can. That means helping others in the community, advancing society, and flourishing at the best.

And humanists are the ones who find value in themselves and one another and respect the personhood and dignity of fellow human beings, not because they are made in the image of something else (human beings are a product of evolution, not the product of a divine plan) but because of what we are—a sentient, feeling species with value and dignity inherent in each person. There is no reason to believe that "meaning" has to come from a supreme being. There is no divine plan or purpose, the humanist recognizes, but we make our own purposes, tell our own stories, and set our own goals. This gives life meaning.[113]

Prof. David Pollock has reiterated humanism as a worldview or *life stance*—a word introduced by the British Humanist Association. The word was coined to fill a gap in the belief when someone does not associate with a particular religion. For example, people with nonreligious beliefs were, and often still are, thought to have "god-shaped holes" in their lives. They were searching for a religion that fits them in their life stance and had nothing to put in its place. There was no perfect word to cover the whole spectrum of fundamental beliefs about what are sometimes called ultimate questions about life, the universe, and everything about the nature of existence and how we should behave. The word *life stance* covers the whole spectrum, he said. Humanism encompasses the features of life stance. But humanism is different from almost all other religions and beliefs because humanism is not an ism in the sense of a body of more or less unquestionable doctrine. One doesn't "convert" to humanism like one converts to a religion; rather, one automatically believes in humanism. It has no sacred texts, no sourcebook of unquestionable rules or doctrine, no liturgy, no founding figurehead, and no structure of authority. Humanism has certain ranges of belief and values, though. You are a humanist to the extent that you do or do not share these beliefs and attitudes, but you are more or less a humanist. Collectively, humanism is a set of beliefs and values that constitute a view of the world, a philosophy based on which many people live their lives. Humanism is an alternative to religion that fulfills much the same function as a religion. It gives us our bearings in the world. It is demanding but immensely rewarding—it puts a lot of responsibility on you to think for yourself, but it provides you with the freedom to do so and a basis on which to make ethical decisions. You don't have to have a religion, but if you don't have a life stance of any kind, you are rootless, without answers or purpose. Humanism provides the answers for those who can't accept religion but want an ethical approach to life, and it brings with it the inheritance of a glorious history and the promise of a better future.[114]

Humanism can be seen as a unity, as a valid concept. Though the name is recent, there is a long tradition—older than any of the main world religions—the nonreligious philosophy of life that we call humanism. Religions have something in common—that is, a belief in god or gods. Yet there are exceptions—not all religions have gods. For example, classical Buddhism, Sikhism, and Jainism have no gods, but they believe in a hidden celestial realm of existence to which their followers aspire. In contrast, humanists' beliefs are naturalistic. They believe that the universe can be explained by natural laws; many of the laws have already been discovered, and many of them are yet to be discovered. Humanists have no belief in an afterlife, and death is the end and there is no survival after death. Almost all religions believe in a continued existence after this life; some of them, a reincarnation in this world; others, a translation to a different realm of existence; and sometimes, they believe in existences before this life too. Humanists think that we are not soul trapped in a mortal body: what we are resides in our bodies and brains, and bodily death means the end of the vastly intricate system of matter animated by electrochemical impulses that make us. Personality and consciousness cannot survive death because these attributes can only exist in a living body; without a body, they do not exist.

Humanists also believe that human beings are moral creatures (with the exception of a few psychopaths and severely autistic people), have the capacity to think in moral terms, and cannot live without morality, which is ingrained in human nature. Humanists say that biology and culture have created our moral sense. There are prosocial behaviors (such as altruism and cooperation) that are necessary for living together with others of your own species. These behaviors are the evolved mechanisms shared by all human beings. Humans have lived as social animals since millions of years before they were even fully human, and all social animals have rules and patterns of behavior that enable them to live together in a harmonious and productive manner. Undoubtedly, if they had not had such rules, they would not have survived on earth. Humans survived since then and made progress with the development of language and the ability for abstract thought. They refined unwritten rules into moral values. Our instincts (which are in our building blocks) are the basis on which the concept of morality is built. Yes, we are not naturally exclusively good, as some instincts are aggressive or selfish, and some are group focused, which are hostile to outsiders. But human nature has the attribute of plasticity; this is because of wrong education, experience, and even of environment. Given the conditions, many people can adopt very antisocial behaviors and indulge in doing wrong things. These suggest that our moral views undergo redesigned process, which is built on by culture, but at the root, morality resides in human nature, hardwired into us. I have discussed various aspects of humanism in earlier sections, so instead of repeating them, I would better highlight one important matter relevant to the tenet of humanism—the meaning and purpose of life.

Life as a whole, as a phenomenon on earth, has no purpose. Human life has no purpose in a sense analogous to any object on earth, which has some purpose. We, as humans, have the capacity to create meaning and purpose for ourselves. What meaning we have is the fact that of our own making; that is, meaning and purpose are human constructs. We only give our own life a purpose; that is, we adopt certain goals that seem meaningful to pursue, and we shape our lives according to our desired goals and work hard to achieve them. At certain turns or cycles of our lives, we assess how much we have succeeded or failed and see whether we have made good use of our time and effort or not. There is common hearsay that good works have good results and bad works have bad results. People talk about your works and put their value judgments about you and your works. For example, my works contribute to those of other people. This is more so after my death in the sense that people will remember me and be influenced by my good works. We also see that at the funeral, we remember someone we have lost and celebrate his life and find comfort in our shared feelings of not just loss but also gratitude for what we have gained from his or her life. As a matter of fact, a meaningful and purposeful life constitutes one's individual way of defining life, which may differ from others. This is because each person is inherently different in their talents, thinking, ideas, learning, and interests. So the purpose and meaning of life differ from person to person. However, one may argue that the very purpose of life is to lead a good, fulfilled life and seek happiness. Here we need to know what happiness is. Happiness is not only the absence of suffering. Everyone experiences every taste life has to offer, such as triumph, despair, joy, hatred, and love.

What makes our lives meaningful and makes us happy? This seems more like personal happiness, which is, by nature, self-centered. Dalai Lama argued that happy people, in contrast to personal happiness, are found to be more sociable, flexible, and creative, and are able to tolerate life's daily frustrations more easily than unhappy people. And most important is that they are found to be more loving and forgiving than unhappy people. Robert Ingersoll, the American freethinker, also contemplated happiness in the way by saying that reason, observation, and experience have taught us that happiness is the only good, that the time to be happy is now, and that the way to be happy is to make others so. David Pollock, while explaining this issue, has mentioned that happiness is something much more substantial. Happiness is something about one's relationships with other people, and our personal happiness is inextricably tied up with that of others and is related to emotional contact. He further has cited another formulation coming from Bertrand Russell: "A good life is one inspired by love and guided by knowledge. These two categories, love and knowledge, recur in humanist conceptions of the good life—love, because one's inner life, one's emotional life is vital and because it is the relational aspect of life, emotional fulfillment, sympathies, and affections, in relation both to others and to the natural world; and knowledge, because the life of the mind, finding things out, learning, knowing, and understanding give joy and fulfillment."

From all these perspectives, we can argue that humanism is an alternative to religion because it fulfills much the same function religions do. Humanism gives us every alternative to religion to lead our lives individually, socially, and globally. It puts a lot of responsibility on us to think of our personal lives as well for others and also gives us the freedom to do so and provides a strong basis on which we can make ethical decisions. We don't have to be associated with a religion that follows rules and practices rituals. However, we need to have a life stance, a guideline without which we are rootless and our lives will be chaotic and lack direction. I am fully convinced that humanism provides the guidelines and answers for those who cannot accept, in principle, any of the religions but want an ethical, moral, evidenced-based, and practical approach to life. Humanism can bring with it the inheritance of a glorious history, which mankind practiced since the inception of modern man. Humanism promises a better future for mankind.[115]

Meaning and Purpose of Life to Humanists

In true sense of the phrase "purpose of life," it can be said that the purpose of life is to live. Resonating with this verse, Dalai Lama adds the word *happy*. So the complete phase stands as "The purpose of life is to live happily." Metaphysically, one may argue, "Life is without meaning." However, we bring the meaning to it. The meaning of life is whatever we ascribe it to be. The naturalistic view seems, to me, a realistic one—"Being alive is the meaning." This is a kind of objective sense of the purpose of life. As humanists, we create our purposes in life, and then striving to achieve them can, in itself, provide a sense of meaning. This meaning is derived from the good we do, such as relationship building, the quest for intellectual growth, the effort for physical and mental development, the satisfaction of productive work, the enjoyment of creative or artistic pursuits, the influence we create within society, working and helping others when needed, etc. But the meaning of these achievements is not driven by any outside authority; rather, we define the meaning of life by ourselves. We can summarize this meaning as "The purpose of life is to live a life of purpose." Obviously, this statement is subjective; one has to find meaning and purpose of his or her life self-defined. "Dum vivimus, vivamus, Horace" (Since we are living, let us live well).

Throughout history, humans have considered the fundamental philosophical question regarding the meaning and purpose of life in the context of the ability or our brainpower, the development of conscious self-awareness, and a higher cognitive ability. As this question faces so many challenges and obstacles, many people have tried to resolve it through religious interpretations. Yet many have suggested that the principal aim of life, in most rudimentary form, is to survive to the fullest efficacy possible and to replicate one's own genetic lineage. However, the meaning of life is open to interpretation in different ways by different scholasticism, ideologies, knowledge, and education of the individual. So in a more subjective sense, the meaning and purpose of life is infinitely open to interpretation.

Secular humanists tend to find the meaning of life to be the pursuit of life in abundance, happiness, pleasure, and love. Categorically, the meaning of life is composed of a number of categories such as psychological, biological, social, political, scientific, and philosophical. We can look at it ourselves through discovering who we really are and then relate the meaning of life with understanding and awareness of one's self. We must realize and pursue the best of our abilities by cultivating individual personality, strength, intelligence, knowledge, skepticism, and empathy. The meaning and purpose of life as humanists in the social, we need to work toward the remediation and reconciliation of social ills and conflicts, to help create a peaceful, cohesive, social inclusion, assimilation, integration, and tranquil social environment. To attain this, we must put our best effort to help others realize and attain their highest potential.

The meaning and purpose of life, in secular humanist's point of view, is to affirm universal human rights and decency and to work toward creating a peaceful and harmonious life among populations worldwide. It is our moral responsibility and obligation to uphold the Universal Declaration of Human Rights, the ideals of democracy, freedom, and the free and open society. It is our utmost responsibility and obligation to work collectively toward increasing educational standards, literacy, cultural enrichment, gender equality and decreasing income disparity between male and female.

Humanism is a philosophy, worldview, or life stance based on naturalism. Therefore, the meaning and purpose of life for humanists, in the context of science and philosophy, is to better understand them and work toward a comprehensive knowledge about the philosophy of life, the universe, and the discoveries based on scientific methods, which we experience every day as they significantly affect virtually every aspect of our lives. Advances in scientific medicine, modern techniques of surgery, anesthesia, pharmacology, and biogenetic engineering have tremendously improved our prospects for a happier, healthier, longer, and more fulfilling life. Scientific research has and will continue to advance our knowledge of the universe and our place within it. So it is the meaning and purpose of life to pursue these ends. So coming back to the point stated earlier, the meaning and purpose of life as humans is not only to survive and replicate but also to derive pleasure and happiness while doing so. It is to love life in its abundance and to share love and happiness with others.[116]

Types of Humanism

Religious Humanism
Any religious belief system that incorporates humanistic beliefs and principles are described as religious humanism. Religious humanism shares with other types of humanism. The basic principles of religious humanism override with humanity—that is, the needs and desires of human beings and the essential components of human experiences. For religious humanists, it is the human and the humane that are the focus of ethical attention. Religious humanists treat their humanism in a religious manner. This requires defining religion from a functional perspective, which means identifying certain psychological or social functions of religion as distinguishing a religion from other belief systems.

Though practitioners of religious humanism did not officially organize under the name of "humanism" until the late nineteenth and early twentieth centuries, nontheistic religions, in conjunction with human-centered ethical philosophy, began date to the Enlightenment era. The Cult of Reason (*Culte de la Raison*) was a religion based on deism devised during the French Revolution by Jacques Hebert, Pierre Gaspard Chaumette, and their supporters. In 1793, during the French Revolution, the cathedral Notre Dame de Paris was turned into a Temple of Reason, and for a time, Lady Liberty replaced the Virgin Mary on several altars. In the 1850s, Auguste Comte, the father of sociology, founded positivism, a "religion of humanity." Auguste Comte was a student and secretary for Claude Henri de Rouvroy, Comte de Saint-Simon, the father of French socialism. Auguste Comte coined the term *altruism*.

One of the earliest forerunners of contemporary chartered humanist organizations was the Humanistic Religious Association formed in 1853 in London. This early group was democratically organized with male and female members participating in the election of the leadership and promoted knowledge of the sciences, philosophy, and the arts. Of the thirty-four original signers of the first *Humanist Manifesto*, thirteen were Unitarian ministers, one was a liberal rabbi, and two were Ethical Culture leaders. Indeed, the very creation of the document was initiated by three of the Unitarian ministers. The Ethical Culture movement was founded in 1876. The founder, Felix Adler, a former member of the Free Religious Association, considered the Ethical Culture as a new religion that rejected the unscientific dogmas of traditional religions while retaining the ethical message at the heart of all religions. Adler believed that traditional religions would ultimately prove to be incompatible with a scientific worldview. He also felt that the vital aspects of religion are important, as religions provide vital functions to encourage good works and religions teach important truths about the world, though these truths are expressed through metaphors that are not always suited to modern understandings of the world. For example, monotheistic religions are based on a metaphor of an authoritarian monarchy, whereas democratic relationships are now understood to be the ideal system. At the beginning, Ethical Culture allowed very little of the ceremony and ritual. Instead, Ethical Culture was religious in the sense of playing a role in people's lives and addressing socioreligious issues. Some Ethical societies subsequently added a degree of ritual as a means of marking special times or providing a tangible reminder of humanistic ideals.

The integration of humanism into religion existed in America in an ideological sense for a very long time. The Free Religious Association (FRA) was formed in 1867, and other less radical groups of early American

Protestants such as the Unitarians and Quakers had existed from the very beginning of the Europeans in the Western Hemisphere. In 1929, Charles Francis Potter founded the First Humanist Society of New York, whose advisory board included Julian Huxley, John Dewey, Albert Einstein, and Thomas Mann. Potter was a minister from the Unitarian tradition, and in 1930, he and his wife, Clara Cook Potter, published *Humanism: A New Religion*. Throughout the 1930s, Potter was a well-known advocate of women's rights, access to birth control, "civil divorce laws," and an end to capital punishment. *Humanist Manifesto I* was written in 1933 primarily by Raymond Bragg and was published with thirty-four signatories. The first manifesto explained of a new religion and referred to humanism as a religious movement meant to transcend and replace previous deity-based religions. The manifesto outlined a fifteen-point belief system, which opposes an acquisitive and profit-motivated society, and emphasized a worldwide egalitarian society based on voluntary mutual cooperation. It is to be noted that the Fellowship of Humanity was founded in 1935 by Reverend A. D. Faupel as one of the humanist churches established in the early twentieth century as part of the American Religious Humanism movement. It was the only such organization to survive into the twenty-first century and is the first and oldest affiliate of the American Humanist Association. American Religious Humanist organizations that have survived into the twenty-first century include the HUUmanists (formerly the Friends of Religious Humanism) and the Humanist Society (formerly the Humanist Society of Friends).

Humanist versions of major religions existed in the major religions in the past such as Christian humanism and Jewish humanism and in many Indian religions like Hinduism, Buddhism, and other belief systems like Confucianism, Taoism, Shenism, and Zoroastrianism, which focus on human nature and action more than religion and theology. However, secular humanists and revealed religious humanists differ in their definition of religion and their positions on supernatural beliefs. They can also diverge in practice since religious humanists endorse religious ceremonies, rituals, and rites. The most popular humanistic approach has been observed in the humanism approach to Buddhism, which shares the fundamental principle of evaluating natural human values. An early exponent, U Dhammaloka (an Irish-born migrant worker in Burma who converted to a Buddhist monk) brought into light the Western free thought and atheist positions with Burmese ritual practice. While most Buddhist groups are more or less humanistic, there is a particular modern Chinese Buddhist group known as *humanistic Buddhism*. The teachings of the modern Chinese Buddhist thought of humanistic Buddhism encompass all the Buddhist teachings of Gautama Buddha. The goal of humanistic Buddhism is the *bodhisattva* way, which means to be an energetic, enlightened, and endearing person who strives to help all sentient beings liberate themselves. Humanistic Buddhism focuses more on issues of the world rather than on how to leave the world behind, on caring for the living rather than the dead, on benefiting others rather than benefiting oneself, and on universal salvation rather than salvation for only oneself.[117]

I believe that any religious perspective that is truly relevant to the twenty-first century must have at least these characteristics. First, it must affirm that human beings are an integral part of nature. We are not separate and distinct from the rest of the natural world; we are part and parcel of it. We are related to every living creature, both plant and animal. The elements of which we are composed—carbon, calcium, iron—are the same elements of which the rest of the universe is made. We are not dominant over nature as we once believed but are its stewards and trustees. Second, therefore, any religion for today's world will affirm humankind's responsibility to preserve and sustain the natural world. The future of life on this planet and indeed the survival of the planet itself depend on it. Third, any viable contemporary religion must take seriously the implications for religion of the remarkable discoveries of the modern, natural, and human sciences. The religion of the future should be a religion that learns from science and adapts its teachings accordingly. Fourth, such a religion will recognize the importance of both reason and reverence.

The human ability to think critically and constructively has made possible our many artistic achievements and medical and technological advances, but it is only reverence, understood as feelings of respect and awe, that can save us from the hubris that would destroy all the good we have accomplished. As Paul Woodruff writes in his elegant little book *Reverence*, "Reverence begins in a deep understanding of human limitations." And it is reverence that keeps human beings from acting like gods. It is thus essential to our true humanity. Finally, the religion of the future must affirm those values that help make our lives more fully human. Karen Armstrong said that the religious quest is not about discovering the truth or the meaning of life but about living as intensely as possible here and now. The idea is not to latch onto some superhuman personality or to get to heaven but to discover how to be fully human.

Becoming more fully human involves the transformation of the mind and heart from self-centeredness to a sense of one's self as part of a larger sacred whole and to a deep commitment to the human and natural worlds. It is about transformation from a shallow life of fear, greed, hedonism, and materialism to a meaningful life of love and caring, gratitude and generosity, fairness and equity, joy and hope, and a profound respect for others. As I understand, it is a religious responsibility and a challenge to learn everything we can about human beings and the world in which we live and to think critically and constructively about what we learn. But we are also emotional beings who need to use our feelings in the service of the best that we know. A fully human person has an open mind and a warm heart as well as a social conscience. As Bertrand Russell suggested, the good life is one guided by reason and motivated by love. The foundation of religious humanism must be built in religious naturalism that makes it possible to state firmly a perspective that includes these characteristics to justify as a universal religion for the twenty-first century. Let me close with these words from Carl Sagan: "A religion that stressed the magnificence of the universe as revealed by modern science might be able to draw forth reserves of reverence and awe hardly tapped by the conventional faiths. Sooner or later, such a religion will emerge." I believe it is emerging among religious liberals today.[118]

Secular Humanism

The term *humanism* in general usually refers to secular humanism as a *default meaning*. Secular humanism traces its roots back to the Renaissance humanism during the fourteenth century, which developed a strong anticlerical tradition in which the repressive culture of the medieval church and religious scholasticism had been criticized intensively. This inheritance was developed further during the eighteenth-century Enlightenment, when the issues of independence, free inquiry into matters of state, society, and ethics were strongly emphasized.

Secular humanism is the branch of humanism that rejects theistic religious belief and adherence to belief in the existence of a supernatural world. Secular humanists (who are often scientists and academics) generally believe that following humanist principles leads to secularism (which asserts the right to be free from religious rule and teachings) on the basis that supernatural beliefs cannot be supported using rational arguments, and, therefore, the supernatural aspects of religiously associated activity should be rejected.

In the philosophical context, the *secular* of secular humanism does not give any place to the veneration of things holy and inviolable. Acceptance of humanist principles lies in a rational consideration of their values and appropriateness without having any sense of their divine origin or of their being worthy of some form of worship. Council for Secular Humanism defines that the secular humanism is a comprehensive, nonreligious life stance incorporating a naturalistic philosophy, a cosmic outlook rooted in science, and an ethical system. Each of these elements has been explained in the following ways:

Secular humanism is comprehensive, touching every aspect of life including issues of values, meaning, and identity. It is, therefore, broader than atheism (which concerns only the nonexistence of God or the supernatural). Rather, secular humanism addresses a lot more of life.

Secular humanism is nonreligious, espousing no belief in a realm or beings imagined to transcend ordinary experience. Secular humanism is a life stance, or what Council for Secular Humanism founder Paul Kurtz has called a eupraxsophy, a body of principles suitable for orienting a complete human life. As a secular life stance, secular humanism incorporates the Enlightenment principle of individualism, which celebrates emancipating the individual from traditional controls by family, church, and state, increasingly empowering each one to set the terms of his or her own life.

Secular humanism is philosophically naturalistic, which emphasizes the fact that nature (i.e., the world of everyday physical experience) is all that we have and that reliable knowledge can be obtained when we query nature using the scientific method. Naturalism states that supernatural entities like God do not exist and warns us that knowledge gained without serious quest of the natural world and unbiased review by multiple observers is not reliable. Secular humanism provides a cosmic outlook, a worldview. What it means is that the basic training and guidance of our lives must be led in the context of our universe, relying on methods demonstrated by science. Secular humanists consider themselves as not being designed and intended human beings by some creator but rather arising through evolution, possessing unique attributes of self-awareness and moral agency.

Secular humanists hold the idea that ethics is consequential, which would be judged by results. This is in contrast to so-called ethics commanded by God, in which right and wrong are defined in advance and attributed to divine authority. "No god will save us," declared *Humanist*

Manifesto II (1973). "We must save ourselves." Secular humanists seek to develop and improve their ethical principles by examining the results they yield in the lives of real people.[119] (See annexure for set of principles of secular humanists.)

A more complete discussion of the secular humanism worldview can be found in David Noebel's *Understanding the Times*. He has discussed in detail the humanism's approach to each of the disciplines such as theology, philosophy, ethics, biology, psychology, sociology, law, politics, economics, and history.[120]

Neo-Humanism

Humanism is going to transform the modern world. The term *neo-humanism* is introduced to present a new approach for dealing with common human problems. It expresses the ideas and values with renewed confidence in the ability of human beings to solve their problems as they encounter and to conquer uncharted frontiers. In our planetary community, the global village, we have the opportunity to peacefully and cooperatively resolve any differences we have. The use of the term *community* is important because of the emergence of global consciousness and the widespread recognition of our interdependence among communities of nations. The worldwide Internet has made communication virtually instantaneous and easily accessible, so whatever happens to anyone anywhere in the world may affect everyone everywhere.

While most decisions that concern human beings are made by them on the local or national level, many issues can transcend these jurisdictions, such as any emergency concerns like regional wars and gross violations of human rights as well as sustainable developments like new ideas in science, ethics, moral values, and philosophy. In the twenty-first century, we inhabit in a global village a common planetary environment where activities in any one country may spill over to others, such as resource depletion and the pollution of the atmosphere and waterways. Of particular importance is the phenomenon of global warming that affects everyone on the planet. Similarly, the possible outbreak of an epidemic or plague (such as the swine flu, tuberculosis, Ebola, wide-reaching malaria, etc.) can have global consequences. So it is essential to coordinate activities such as production and distribution of vaccines, application of common quarantine policies, and so forth.

Many other key issues are of serious importance to the global community, which require cooperative action, such as the preservation of unique species and ecosystems, prevention of excessive fishing on the high seas, management of economic recessions, development of new technologies and protection of cyber threat with promises for humankind, amelioration of poverty and hunger, reduction of great disparities in wealth, seizing the opportunities to reduce illiteracy, addressing the need for capital investments or technical assistance in rural areas and depressed urban centers, providing for public sanitation systems and freshwater, and sharing the earth's water resources for all mankind. Also important is the need to liberate women from ancient repressive social systems and religious customs and attitudes and to emancipate minorities, such as the untouchables in India, who suffer from religious prejudice and caste systems. Similarly, gays and other sexual minorities need to be liberated wherever they suffer harsh punishment, because of their sexual orientations. Of particular importance of the concern is that an ongoing broad range of campaign for women's education and social status improvement is vital. The list of indignities, neglects, and disparities is huge.

Undoubtedly, we agree that science and technology must be used for the service of humanity. In a rapidly changing world, fresh thinking is required to move civilization forward. We, as superior species on earth, have serious concern with reconstructing old habits and attitudes to make peaceful coexistence with happiness and well-being within this global village where every person can realize the best of the good life for self and can share with others.

There are various forms of religious belief systems in the world. One can argue that religions (which vary in terms of degree among them) stand in the way of human progress and development. On the contrary, neo-humanism aims to provide a good philosophy for those who are skeptical of the traditional forms of religious belief yet maintain that there is a critical need to bring together the varieties of belief and unbelief and provide a positive outlook for the benefit of the global community. Commonly, believers include all the major world religions (i.e., Christianity, Islam, Judaism, Hinduism, Confucianism, Taoism, Shinto, Sikhism, some forms of Buddhism, etc.) and also the many denominations within each. There are over four thousand religions or faith groups, ranging from dogmatic extremists to religious humanists. But the fact is that most religious belief systems are so deeply entrenched and rooted in faith and unchanged tradition that it is difficult to be receptive to the new ideas and open dialogue. Historically, believers have often attempted to suppress dissent and persecute those who do not believe religious doctrines acquired at birth. We witness the conflicts between Protestants and Roman Catholics, Sunni and Shiites, Hindus and Muslims, which, continuing to this day, have erupted into violence.

Nonbelievers, on the other hand, a relatively smaller group who focus primarily on the lack of scientific evidence for belief in God and the harm, often committed in the name of religion. The new atheists or the secular atheists have been very vocal, claiming that the public has not been sufficiently exposed to the case against God. In fact, the community of religious dissenters includes not only atheists but also secular and religious humanists, agnostics, skeptics, freethinkers, and even a significant number of religiously affiliated individuals. The latter group may be only nominal members of their congregations, and they seldom attend church, synagogue, temple, or mosque primarily for social reasons or because of ethnic loyalty to the faiths of their forebearers, but they do not accept the traditional creeds and rituals. Interestingly, ethnic identities can be very difficult to overcome and linger long after belief is faded; it may take many generations. Although many of such individuals may be skeptical about the creed or rituals, they may still believe that without religion, the moral order and rule of society may be likely to collapse.

Religious identity is instilled in childhood, at the very earliest ages when a person is being defined. It is so difficult to say that one is no longer a Christian, Muslim, Jew, or Hindu even though one rejects the religion and no longer believes in its doctrines and creeds. It is conceivable that religion is not only a set of beliefs but also a way of life, a commitment to religious cultural traditions and institutionalized moral practices and rituals. When critics of religion only focus on its beliefs, which are taken literally, many believers interpret and analyze them metaphorically or symbolically and judge them functionally for the needs that they appear to satisfy the believers. One may argue that the strongest case against religions is that they are often irrelevant to the genuine practical solution of the problems faced by individuals or societies. We know that the major religions were rooted in ancient premodern nomadic or agriculture-based cultures that are no longer applicable to the urban, industrial, and technological global civilization that has emerged in modern time. We may argue that not enough attention has been paid to humanism as an alternative to religion. But it is important to know that humanism presents a set of principles and values that began during the Renaissance and came to fulfillment during the modern era. It marked a turning point from the medieval thoughts of divine order and salvation to a modern philosophy of this life here on earth, the quest for meaning, value, and the good life and social justice in modern democracies and economies that served people's demands, tastes, and satisfactions.[121]

On January 30, the International Humanist and Ethical Union (IHEU) and the European Humanist Federation (EHF) announced a new project to help abolish blasphemy laws around the world. The "end blasphemy" laws campaign is thought to be the first campaign focusing solely on the issue of laws against "blasphemy," including "ridicule" and "insult" to religion or "hurting religious sentiments." The coalition behind the campaign, led by the International Humanist and Ethical Union (IHEU) and the European Humanist Federation (EHF) and numerous coalition partners, currently represents around two hundred humanist and secular organizations globally and is open to all groups who oppose "blasphemy" laws, including religious and secular communities, human rights groups, and all advocates of freedom of expression. Sonja Eggerickx, president of the IHEU, said, "In the wake of the *Charlie Hebdo* killings, there have been renewed calls to abolish 'blasphemy' and related laws in almost every country where they still exist."[122]

To know more about the neo-humanism approach and statements, we recommend reading the article "Neo-Humanist Statement of Secular Principles and Values: Personal, Progressive, and Planetary" by Paul Kruz.[123]

Humanism in Perspective

We Are Human Beings
Human beings are neither entirely unique from other forms of life nor final product of some planned scheme of development. The available evidence shows that humans are made from the same building blocks of which other life-forms are made and are subject to the same sorts of natural pressures. All life-forms are constructed from the same basic elements and the same sorts of atoms as of nonliving substances. These atoms are made of subatomic particles that have been recycled through many cosmic events before becoming part of us or our world. Humans are the current result of a long series of natural evolutionary changes but not the only result or the one final event. Continuous change can be expected to affect ourselves, other life-forms, and the cosmos as a whole.

There is no compelling evidence to justify the belief that the human mind is distinct and separable from the human brain, which itself is a part of the body. All that we know about the personality indicates that every part of it is subject to change caused by physical disease, injury, and death. Thus, there are insufficient grounds for belief in a soul or some form of afterlife. The basic motivations that determine our values are ultimately rooted in our biology and early experiences. This is because our values are based upon our needs, interests, and desires that often relate themselves to the survival of our species. As humans, we are capable of coming to agreement on basic values because we most often share the same needs, interests, and desires and because we share the same planetary environment. It is possible to develop a system of ethics based upon basic human needs, drives, motivations, and characteristics when reason and empathy are consistently applied toward the meeting of human needs and the development of human capacities. In the meantime, human ethics, laws, social systems, and religions will remain a part of the ongoing trial-and-error efforts of humans to discover better ways to live. When we are left free to pursue our own interests and goals, to think and speak for ourselves, to develop our abilities, and to operate in a social setting that promotes liberty, we can do better toward the goal of greater self-understanding, better laws, better institutions, and a good life.

Humanists deserve a society that must be based on social policies conducive to the general population who are committed to free inquiry and who see the value of social systems that promote liberty; we encourage the development of individual autonomy. In this context, we support such freedoms and rights as religious liberty, church-state separation, freedom of speech and the press, freedom of association (including sexual freedom, the right to marriage and divorce, and the right to alternative family structures), the right to birth control and abortion, and the right to voluntary euthanasia. Humanists support the laws that protect the innocent, deal effectively with the guilty, and secure the survival of the needy. They want a system of criminal justice that is swift and fair, ignoring neither the perpetrator of the crime nor the victim, and considering deterrence, restoration, and rehabilitation in the goals of penalization. However, not all crimes or disputes between people must be settled by courts of law. A different approach involving conflict mediation, wherein opposing parties come to mutual agreements, also has the support of humanists.

As humanists who see potential in people at all levels of society, we encourage an extension of participatory democracy so that decision-making becomes more decentralized and involves more people. We look forward to widespread participation in the decision-making process in areas such as the family, the school, the workplace, institutions, and government. In this context, we see no place for prejudice on the basis of race, nationality, color, sex, sexual orientation, gender identification, age, political persuasion, religion, or philosophy. And we see every basis for the promotion of equal opportunity in the economy and in universal education.

As humanists who realize that all humans share common needs in a common planetary environment, we support the current trend toward more global consciousness. We realize that effective environmental programs require international cooperation. We know that only international negotiation toward arms reduction will make the world secure from the threat of thermonuclear or biological war. We see the necessity for worldwide education on population growth control as a means toward securing a comfortable place for everyone. And we perceive the value in international communication and exchange of information, whether that communication and exchange involve political ideas, ideological viewpoints, science, technology, culture, or the arts.

As humanists who value human creativity and human reason and who have seen the benefits of science and technology, we are decidedly willing to take part in the new scientific and technological developments around us. We are encouraged, rather than fearful, about biotechnology, alternative energy, and information technology, and we recognize that attempts to reject these developments or to prevent their wide application will not stop them. Such efforts will merely place them in the hands of other people or nations for their exploitation. To exercise our moral influence on new technologies and to have our voices heard, we must take part in these revolutions as they occur. As humanists who see life and human history as a great adventure, we seek new worlds to explore, new facts to uncover, new avenues for artistic expression, new solutions to old problems, and new feelings to experience. We sometimes feel driven in our quest, and it is the participation in this quest that gives our lives meaning and makes discoveries possible. Our goals as a species are open-ended, and we will never stop in the quest of new things.

One of the most original contributions to Chinese thought is the view of Xunzi (a philosopher and pragmatist) that turns human endeavor into naturalistic view rather than supernatural.

Instead of regarding Heaven or Nature as great and admiring it, why not foster it as a thing and regulate it? Instead of obeying Heaven and singing praise to it, why not control the Mandate of Heaven and use it? Instead of looking on the seasons and waiting for them, why not respond to them and make use of them? Instead of letting things multiply by themselves, why not exercise your ability to transform them? Instead of thinking of things as things, why not attend to them so that you don't lose them? Instead of admiring how things come into being, why not do something to bring them to full development?

Such view emphasizes the fact we should think of nature as "things" to be explored rather than wait for some heavenly acts. From this view, we get good understanding of harmony between human and nature. This view had profound influence not only on Chinese thought but also in the Enlightenment ideas of the eighteenth- and nineteenth-century Western world (Burns 2006).

Humanism in Global Culture

With no doubt, the most important intellectual task of the present, within the global cultural perspective, is the establishment, implementation, and practicing of a new kind of humanism. Current global conflicts in politics, economics, social, culture, and religion demand strongly for defining and strengthening a global culture of values, morals, ethics, and humanity. Thus, a new role of humanity in the context of human belief system needs to be redefined. Extremism, fundamentalism, and terrorism in all aspects of human life (religious in particular) as well as hunger, poverty, misery, and economic disparity between rich and poor globally provide sufficient evidence for the necessity of redefining humanity. With no equilibrium system in world economic and social structures, diversified religious doctrine, and the absence of consensus and commitment to implement the principles of good governance in many countries of the world, plus powerful societal tensions, sectarian and religious conflicts require new and better understanding and actions for a new humanistic order. We share a common nature, and this nature includes the mental formations of culture, which gives sense and meaning to human life and self-awareness as humans. We face challenges posed by globalization. In the area of cultural orientation, two options may be available for us— the one would be a clash of civilizations, and the other could be the establishment of a new transcultural ethos of mutual recognition and understanding among each of the cultures based on the shared norms and values. In the West, humanism has always been a central issue. However, some credibility has been lost after postmodern and postcolonial Western critiques. Now there is a pressing need for newly viable concepts of what it means to be human and humanity at large.

Religion has always played an important role in the formation of cultural identities. But how much could it do so universally? As observed, religion seems to be a worrisome obstacle to the development of a common value system that has universal appeal. One may argue that religious fundamentalism strongly opposes any attempt to establish a universalistic humanism as a leading principle of understanding of cultural diversity and to build a harmonious relationship among them. Why? Because the humanist outlook goes along with a secular way of life, which is not the way religion recommends. But even beyond this opposition between religion and secularism, religion remains a problem of intercultural communication. When religious belief combines with universally accepted truth, it negates all other belief systems as untrue and invalidates other religious views to be valid. Perhaps the best solution is to overcome the specific forms of static religious doctrine in favor of a universally accepted common morality or system of ethics and values. But religion cannot be reduced to or even be dissolved into that morality. Rather, it remains, alongside its own logic, within the domain of specific cultural and religious orientation.[124]

Help Humanity by Adopting Humanism

> *You must be the change you wish to see in the world.*
> —Mahatma Gandhi

When forced to summarize the general theory of relativity in one sentence: Time and space and gravitation have no separate existence from matter . . . *Physical objects are not in space, but these objects are spatially extended.* In this way the concept "empty space" loses its meaning . . . The particle can only appear as a limited region in space in which the field strength or the energy density is particularly high . . . The *free, unhampered exchange of ideas and scientific conclusions is necessary for the sound development of science, as it is in all spheres of cultural life* . . . We must not conceal from ourselves that no improvement in the present depressing situation is possible without a severe struggle; for the handful of those who are really determined to do something is minute in comparison with the mass of the lukewarm and the misguided . . . *Humanity is going to need a substantially new way of thinking if it is to survive!* (Albert Einstein)

Our world is in great trouble because of human behavior founded on myths and customs that are causing the destruction of nature and climate change. We can now deduce the most simple science theory of reality—the wave structure of matter in space. By understanding how we and everything around us are interconnected in space, we can then find solutions to the fundamental problems of human knowledge in physics, philosophy, metaphysics, theology, education, health, evolution, ecology, politics, and society. This is the profound new way of thinking that Einstein realized: we exist as spatially extended structures of the universe—the discrete body an illusion. This simply confirms the intuitions of the ancient philosophers and mystics. Given the current censorship in physics/philosophy of science journals (based on the standard model of particle physics / big bang cosmology), the Internet is the best hope for getting new knowledge known to the world. But that depends on us; the people who care about science and society realize the importance of truth and reality.

Let us think this way: seven billion people on earth believe in one common thing—humanity. But all do not belong to one religion. Rather, they belong to different religions. Also, the fact that humanity is the key theme of all religions means that humanity is the backbone or lifeblood of all religions. So humanity is the religion of all religions. Is it not better to be associated with the religion of religions—humanity—instead of associating with one component of humanity that is one specific religion? The diagram below explains.

HUMANITY					
Hinduism	Buddhism	Judaism	Christianity	Islam	All Others

By associating with one universal religion (humanity), seven billion people can breed love among them, whereas different religions breed hate, animosity, hostility, riots, and even war among one another. If humanity unites people of all religions in one platform, then different religions divide people among religions. Humanity, with the quality of being humble, brings inner peace. The less compelled you are to try to prove yourself to others, the easier it is to feel inner peace. Morality comes from humanism as it fights for human rights, dignity, progress, and rationalism, but religions of varied ideologies steal morality for their purposes. We don't need a religion to have morals; if we can't determine right from wrong, then we look for empathy, not religion. Humanism is important because of the fact that having a nonsuperstitious worldview allows us to make more ethical choices based on the general desire to do the most possible good. In terms of the reality of the world today, Dalai Lama has said that all the world's major religions, with their emphasis on love, compassion, patience, tolerance, and forgiveness that can promote good values, do not appear to be adequate in recent time. He also opined that the grounding ethics in religion is not too good to appreciate.

Looking at the fact of what the function of humanism serves in the society, it may be called another religion (in the sense of universal religion). It is a worldview that gives answer to the questions about human beings such as the place of humans in the universe, relations to one another and to other creatures on earth, and many more at the microlevel of human lives. Humanism is a basis for human values, ethics, and morality. However, there are differences between humanism and other forms of religion, as there are also differences between religions, as we all know. There is another important difference between religion and humanism, and that is the role of science. I would consider thinking the fact that humanism is superior to religion in the sense that humanism accepts established outcomes of science. In this respect, humanism is not an ideology, because if something turns out to be a fact, then it will be accepted. The ultimate truth as written in the religious books can't accept as it rather should be open to revision.

What is very important to look at in terms of values and ethics of life is that humanism is a progressive philosophy of life that, without religious dogmas and supernatural beliefs, our ability, human power, and responsibility to lead ethical lives of personal fulfillment and collective welfare, aspire to the greater good of humanity. It is the study and knowledge of what it means to be a good human being. It is not about dogma or absolute prescriptions for living our lives; rather, it is about striving to do good and better. Bertrand Russell said, "A good world needs knowledge, kindness, and courage. It does not need a regretful hankering after the pastor, a fettering of the free intelligence by the words uttered long ago by ignorance." Let us refresh our thinking by the following comparisons:

"Reason, not superstitions; ethics, not dogma; respect, not worship; courage, not fear; morality, not religion; clarity, not delusion; good, not God; skeptic, not cynic; pragmatism, not ideology." How do they sound? They sound incredibly good to me.

Humanism emphasizes the human concerns as the basic principle. It overcomes tradition, religious dogma, or creed. Humanists seek to discover what best promotes human flourishing while leaving behind those religious beliefs and practices that would prevent humanity from achieving its full potential. This principle can be expressed in some core values such as reason, compassion, hope, etc. Humanists value reason, or the use of the intellect and practices like the sciences and philosophy, as the best way to generate and achieve accurate knowledge about the world we live in. They reject the supernatural, as promoted by religion, explanations for phenomena. Stephen Hawking said, "Is the way the universe began chosen by God for reasons we can't understand, or was it determined by a law of science? I believe the second. If you like, you can call the laws of science God, but it would not be a personal god that you could meet, ask questions." A similar idea has been promoted by Bertrand Russell, and he said, "Religion is something left over from the infancy of our intelligence. It will fade away as we adopt reason and science as our guidelines."

Humanists are driven by compassion, the idea that all people (regardless of nationality, ethnicity, race, creed, sexual identity, or other characteristics) are fundamentally of equal moral values. They also look to the future in the best hope with the belief that human beings, when they work together, can build a better world. It goes to suggest that humanism, rather than religions of varied ideologies, can and will make a better world with seven billion people. Could we not say humanism is a religion of mankind vis-à-vis universal religion? Many tributaries of thoughts flow into the mainstream of humanist thought, but some are particularly significant. For example, much of modern humanism is inspired by the principles of Enlightenment, meaning that a commitment to reason is a tenet to change society, and a commitment to science is the best way to learn and understand the world and the laws by which the universe operates. The journals, worldwide media networks, debating societies, learned academies, and people at large of the Enlightenment must pave the way for the marketplace of ideas and promote humanism, a concept that characterizes modern culture and thoughts that humanists and associates embrace wholeheartedly.

Humanists believe that people should be free to think and discuss any thought, including religious. Anyone, like humanists, has the right to question any aspects of religion(s) without any fear of persecution or threat of death. Who in the twenty-first century wants to go through this kind of fear prescribed by religion? The other major factor that has significant influence derives from liberal religious movements, including liberal Christian and Jewish movements such as transcendentalism and Unitarian Universalism. Such movements seemed to have significantly de-emphasized the role of God and the supernatural, thinking alike humanism. In fact, many of the signatories of the first *Humanist Manifesto* were religious liberals who found that their questioning of God's role in the cosmos led them to humanism.

It is important to note that the full range of values and ideals central to humanism is not so easy to capture in a short statement or through a brief dialogue. The task is more challenging because humanism is nondogmatic by ideology, concept, and design. There are no required "creeds" or "religious dogma" in which humanists believe. There is no holy book of humanism that lays out what humanists should or should not do or should follow word for word like scriptures or holy books of religions. Like any tradition, humanism has no single founder and has no ultimate authority, but it promotes that ethics is an ever-changing field of human practice that must be altered to fit the context and the times. However, humanists follow a creedal document, which is a set of *Humanist Manifesto*s, a record of consensus view of what humanists believe at a particular time, subject to be revised when circumstances change. Humanism has a very complex relationship with traditional religions. While humanism is not inherently "antireligious" in the sense that humanism asserts that all aspects of religious practice are harmful and inhumane, at the same time, humanism is not inherently "proreligion" either because it does not claim that all elements of religious practice are positive and valuable. Rather, humanism eliminates aspects of religious practice that are found to be harmful, inhumane, and dehumanizing. Humanism is based on those elements that affirm and promote human flourishing, including moral, ethical, and all aspects of character development. The question is how humanists see religions. In response, humanists' optic is that they see religions and religious practices as human created and that they seek to ensure that religions do exist and need to serve human concerns rather than dictate them.

Humanists gather knowledge and get inspiration from various sources from Mother Nature and from the people. To mention a few, Carl Sagan, a humanist, found the boundless beauty of the cosmos with magnificent awe; philosopher Bertrand Russell found the rigors of geometry "dazzling as first love"; Ernestine Rose, a social reformer and activist, found her blissful satisfaction in her work to promote women's right to vote in political elections and abolitionism; Margaret Sanger, founder of Planned

Parenthood, wanted to change attitudes toward reproductive rights; and Gene Roddenberry, creator of *Star Trek*, expressed his humanism with a hopeful vision of human life among the stars; and many more. Nevertheless, humanism lives and excels wherever human beings reach out to better understanding of the universe and our role within it, wherever human concerns are placed above the will of a god or the needs of a tradition or religion and wherever people believe that a better world is possible in *this* life. We, therefore, seek the values and importance of humanism to a wider reach among human mankind.[125]

Is Humanism Important to the Future of Mankind?

Yes, of course, humanism is very important to mankind. Humanism promotes that religion is not adequately designed to serve human concerns and, therefore, should not be allowed to dictate how societies should operate. All religions must operate by human spirit and within a framework of law determined by democratic mandate. Since the patriarchal religions depend on the ideologies so old that they have no relevance to contemporary conditions, they are unable to give any help in response to various problems of the world and its inhabitants. Indeed, fundamentalist adherents to these ideologies deny the validity of science. They are the most dangerous group of people on earth today. Since humanism and secularism are closely related, humanism is vital for our future. We need to change the way we do things so that science and a real understanding of the long-term consequences of what we do become the basis of our decision-making but not the incoherent ramblings of the Dark Age scriptures and holy books or the obsession that they tend to encourage in the short term (Rod Fleming, author of the book *Why Men Made God*). Humanism is a rational approach to problem-solving that is geared toward making the world a better place for humans to live, and it is pretty much the only approach that will help us achieve our potential as a species and give us a chance at avoiding extinction, said Jennifer Hancook, humanist author.[126]

Humanism is a good "check and balance" against organized religion. Humanism is an approach as to how we treat one another and how we conduct our lives. One could say that humanism is a perspective, a concept, a way of life like spirituality, pragmatism, capitalism, rationalism, etc. Without all those different ways of life and the viewpoints, human beings won't be able to build a better human society, a better world. Otherwise, our future is going to be very bleak. No society can thrive based on a single unique and universal perspective or approach to life. Let us think that we have to be a humanist to be human. If one really wants to be a human, one has to realize that collective humanity exists, which is, in fact, divinity, and each human being is a potentially divine one of absolute purity, sense of purpose, and unconditional love reflecting that divinity. This divinity must exist beyond individual dogmas and isms. The approach of Vedanta philosophy is unique in that aspect, as it harmonizes all the beliefs and ideologies, which takes us to the core of divinity (Seshadri Narasimhan, Truth is One Eternal, Perceptions Are Many).

As Prof. Lawrence Krauss said, humanism characterizes the spirit that he has tried to use as a guide in his personal, professional, and public activities. He summed up the spirit as follows: "It is up to us determine the nature of the way in which we carry out our lives, using a combination of reason, intelligence, and compassion. No one is taking care of us but us. Bad decisions produce bad consequences, and we must take responsibility for them, and, if possible, take actions to mediate or alleviate them."[127]

Humanism offers us one of the most important drivers of change that can improve our future. It is an optimistic view that humanism has an important future for human beings. So one may argue that it is possible to imagine a future without the tyranny of religious myth, dogma, and superstition. Perhaps we are not very far away to have the change. Our generations will not see, but future generations will. It is important that while educating our children, we should encourage them to question everything, not be satisfied with blind faith and unsubstantiated claims but be skeptical of a priori beliefs, not only of their own but of their parents' or their teachers' as well, encouraging skeptical thinking and promoting a culture by which questions may be satisfactorily answered with logic, reason, and scientific evidence. This will help prepare them to be good and responsible citizens who can address the demands of a democratic and free society. Evidence suggests that seeds of religious doubt are planted more among the scientifically literate populace, especially among the younger generation. It is always good to be skeptical, especially about ideologies you learn from scriptures and religious figures. Gautama Buddha advised the following:

Do not believe in something because it is reported. Do not believe in something because it has been practiced by generations or becomes a tradition or part of a culture. Do not believe in something because a scripture says it is so. Do not believe in something believing a god has inspired it. Do not believe in something a teacher tells you to. Do not believe in something because the authorities say it is so. Do not believe in hearsay, rumor, speculative opinion, public opinion, or more acceptances to logic and inference alone. Help yourself, accept as completely true only that which is praised by the wise and which you test for yourself and know to be good for yourself and others (The Buddha, the Kalama Sutta, Anguttara Nikaya 3.65, *Sutta Pitaka*, Pali Canon). This sounds like humanism.

Studies suggest that when people, especially at a younger age, are being skeptical to anything doubtful and able to open questions, they could make people better lifelong learners. That, in turn, means that people who perceive their views on evidence rather than faith are likely to be better citizens. This kind of learning through education offers the best opportunity to help people be open thinkers and immunized against the intellectual virus associated with dogma and superstition. When people develop this mind-set, they would be able to publicly accept and promote the fact that many of the claims of the sacred books of the world's major religions are not valid. We must understand the beauty of science—physics, mathematics, and cosmology. If we do so, only then can we question many texts written in the holy books—for example, the sun orbiting the earth, the universe being created in six days, Prophet Muhammad's night journey to meet God, the Virgin Mary being pregnant without having sexual intercourse with a man, and many others. We all know that there are places in the world where one risks decapitation for questioning certain religious claims. However, in a rational world, one can argue that questioning many religious claims, which are dubious in nature, should be viewed as inappropriate. Christopher Hitchens said that religion poisons everything. This can be debated, but with moderate view, one can argue that in the current environment, religion has devastating consequences in the political process in many countries in the world.[128]

Many Christians do not consider their faith as the sole source of wisdom in the world, although they may consider Christianity as the best religion. This understanding, to a great extent, allows the people to learn from other religions as well. The same is true for many Jews, Muslims, Buddhists, and Hindus. It is true that religion does not change, because it claims to be the eternal guide for solutions to the problems of mankind; as such, anything that claims to be eternal does not accept and agree to change. But the followers of religions do. Christianity has not changed, but Christians have changed. Hinduism has not changed, but many Hindus have changed. In the case of Islam, not only has it not changed, but it also was neither expected to change nor had any windows to change. This is the biggest problem in Islam, because Muslims refuse to change. Refusal to change (for any of the religions) makes religions the ultimate goal for all believers who want to maintain the status quo. In fact, religion becomes the instrument of holding on to power in the hands of orthodox rulers and their agents.

I want to reproduce what Prof. Lawrence Krauss asked; would it be naive to imagine we can overcome centuries of religious intransigence in a single generation through gathering knowledge, education, and thinking outside religious box? I think perhaps not. But the stakes are very high not to try. As Feynman warned us, "It is our responsibility to leave the men of the future with a free hand. In the impetuous youth of humanity, we can make grave errors that can stunt our growth for a long time. This we will do if we, so young and ignorant, say we have the answers now, if we suppress all discussion, all criticism, saying, 'This is it, boys! Man is saved!' Thus we can doom man for a long time to the chains of authority, confined to the limits of our present imagination."[129]

Humanists think that the entire field of human development is possible within the human frame, which is a cornerstone of optimistic humanism. It is arguable that some form of religion is probably necessary. However, instead of worshipping supernatural figures and God, it is wise to represent the higher manifestations of human nature in art, love, science, rational thinking, and intellectual demonstration. For many years, humanists and rationalists have worked to help people get rid of superstitious thinking and mean-minded self-interest and tried to bring the light of reason and empathy. During the European Enlightenment, humanist intellectuals were extremely optimistic. By the 1700s, in Europe, science was showing religious myths to be wrong or unnecessary to explain the world.

We humanists can establish hope from the Enlightenment's accomplishments of the past hundreds of years. We can establish hope from the advancement of human reason in solving tough problems of human beings. We can expect that we will continue to cope with the problems of the present and the future. From a humanist perspective, progress has been made in the area of superstitious religious belief. For example, the Unitarian Church officially endorsed nonsupernatural religious beliefs. This kind of reform has to be made in other religions as well.

Humanism over Religion

We have seen that humanism is important to human beings. But the question is, are we, as humanists, taking the place of religion? The answer is yes. Humanism is a progressive liberal philosophy of life of people, which upholds freedom of and freedom from religion. This is a fact. Humanism had struggled so long into existence within global societies by rejecting religious beliefs of all kinds. It emerged from societies where obedience to religious authority was imposed and strongly enforced in the society. It flourished within cultures where scientific evidence about the natural world was not only controlled but also suppressed or denied by the religious authorities when it did not fit the prevailing faith. For thousands of years, it has been witnessed that any skepticism of religion has been responded with hostility by the religious followers. As a result, many humanists have paid a heavy personal and social price for rejecting or publicly challenging the prevailing religious faith. Despite the struggles, humanism has grown rapidly and has made much headway into the societies across the world.

Supernatural entities are not bound by the laws of nature, nor are they available for empirical examination or scientific experimentation. Most of the religions are more or less similar in believing supernatural components, but they are different in the way followers conceive, describe, and practice them. The imaginative impulse that has been perceived by the followers gives rise to thousands of different religious beliefs, which, in most cases, contradict one another. Since the nature of religion is diverse, we strongly admit the fact that the existence of religion is influenced by the culture of the society. Nonetheless, it is not appropriate to say that religious claims are true, real, or rational. On the contrary, humanists reject supernatural entities because of the fact that those entities are supported by valid and reliable evidence. Sadly, this view was not shared by the majority of the religious believers in the past and is not shared now.

To know why humanism is taking over religion, it is important to understand why religion came into being in human minds. We have discussed this issue in the earlier section. However, let us reiterate to review in humanistic perspective. Throughout human history, people have been asking questions about nature, lives on earth, the universe, and so on. Some of the common questions are the following:

- What is the meaning of life?
- Does life have a purpose?
- How does one lead a good life?

- How is morality defined?
- What happens after death?
- How did the universe come into being?
- Who created the universe?
- How should we treat other species?
- Why is there something rather than nothing?
- Why do we exist?
- Why this particular set of laws and not the others?

Zelda Bailey discussed briefly about humanism and religion during an introductory talk to the newly formed East London Humanist Group in September 2012. His explanations seemed logical and rational, at least to humanists' perspective. We know that all creatures, including humans, need to survive and reproduce in the face of innumerable natural odds and, therefore, need to make sense of their natural environment, which is extremely complex and harsh.[130]

In humanists' perspective, the whole idea related to the questions mentioned above is central to making sense based on the tenet that there are no predetermined guidelines, dogmas, or doctrines derived from God or gods. So humanists have to construct convincing and satisfying values, principles, and ideals on their own in the absence of ready-made answers. The beliefs developed in this manner rely on personal responsibility, kindness, the wish to reduce or end fear and not cause suffering, and respect for the rights of others, as they are available by the most accurate knowledge of the world currently available. During this process, the rationalistic view of our nature drives us to question ourselves. This human drive is stronger in some people than in others, but it is a strong force in human beings, freethinkers and humanists in particular. It is this force and drive to know, understand, and respond appropriately to those questions.

In religionists' perspective, on the other hand, many explanations, rules, and pathways are provided ready-made in sacred texts and books, in ancient rituals, and in the declarations of those who claim special knowledge through revelation, such as prophets, saints, monks, and religious thinkers. Thus, it is the acceptance of religious beliefs that constitutes their faith. Religious people insist that these are religious questions and can be answered only by God or a god or other supernatural forces. In humanistic point of view, answers and explanations that rely on God or supernatural causes are irrational and unconvincing. But why are religious answers so widespread and deeply embedded in societies and cultures across the world? The most obvious answer is that they took root many thousands of years ago in a prescientific age when knowledge of the world was limited and information contained in oral accounts was passed on from one generation to another.

As we know, to survive, people deal with a hostile and unpredictable world, people created stories to account for dangerous phenomena, and they devised imaginative ways for dealing with them. In the prehistoric time before writing was invented, oral storytelling was the primary means of transmitting historical knowledge and practical advice. Eventually, many of these ancient stories were written in the religious scriptures and holy books at a later date. Whenever the natural cause of something was not understood, a supernatural god, spirit, or force was imagined to be the actor of that cause. This interpretation appeared to provide certainty in a confusing and frightening world, obviously, without any scientific evidence. By accepting the truth of religious interpretations that could not be disproved and rejected, believers of all religious faith could be reassured, pacified, and comforted. Moreover, it had also the advantage of strengthening group cohesion and solidarity within a community of like-minded people, irrespective of the particular religious faith.

In societies where religious belief, practice, and social expectation out of those beliefs are very strong, it is extraordinarily difficult for an individual to take a contrary path. If someone does so, it is likely to have profound family and community repercussions; even the person is likely to be cast out from the family and community. Anyone who makes the decision to act in this way under these circumstances takes considerable personal courage and strong determination. One could imagine that the fear generated in this circumstance prevents someone from doing so. In other words, this fear acts as a powerful inhibitor. Yes, the power of religion is so strong that it shapes various customs by faith, which affected societies in the past and continues to affect societies now globally. However, many of these religious customs, such as discrimination based on caste, the ritual mutilation of children, the torture and murder of witches, etc., are intolerable and unacceptable in the world in the light of human rights, scientific knowledge, great advances in education, mass communication, and technology. Therefore, humanists embrace pluralism and strongly and vehemently oppose the harms and injustices carried out in the name of religious rituals, traditional rules, obligations, or customs. Humanists also confront the governments that allow them to continue. The voice of humanists is clear and loud—down with human rights violation, torture, and killing of people in the name of religion.

Religious people depend on supernatural explanations to make sense of their lives. On the contrary, humanists depend on reason and experience to understand the world and nature at large. So they emphasize scientific evidence to independently confirm the knowledge across disciplines to provide a rational basis for their approach to life. Obviously, though, it is not possible, even with the scientific methods, to know the entirety of a discipline because science is an ongoing process and knowledge gained by science is based on results, which is provisional and subject to further research and analysis. Yet science does provide us with a sounder foundation for confidence and trust than any religion. This is because of the fact that scientific facts and interpretations about the nature of the universe are tested empirically and objectively to show that they are valid and reliable, unlike religious faith, which claims it is always subjective and impossible to verify scientifically. But for humanists, it is the explanatory power and integrity of science that make us understand the world and nature. The fact is where evidence is strong, faith diminishes. Although debatable, one can argue that humanism is taking over the place of religion in the prevailing modern, civilized societies across the world.

There are other issues in the religious systems of belief than to supernatural knowledge. One of the most important ones is moral behavior. The religious path to morality follows a code of conduct, a set of commandments, or rules revealed by a god or other spiritual authority such as a prophet, a priest, saints, etc. The humanist, on the other hand, finds morality in the natural world through nonreligious ways. Humanists emphasize that the concept of morality has developed over thousands of years of social and cultural evolution, revealing evidence to show that morality development is the natural continuation of the manner in which social relations have evolved in earlier species. While studying the social behavior of many animals in their natural habitat, we find in them the traits such as cooperation, teamwork, fairness, love, altruism, nurturance, loyalty, bravery, curiosity, joy, etc., which are similar to human traits.

Likewise, the evidence of selfishness, jealousy, deceit, greed, anger, aggression, loneliness, misery, etc., can also be observed in other species. These social and emotional propensities have evolved gradually over millions of years and continue as traits in human behavior. As human beings, by our intelligence and superior conscience, we can modify values and morals by experiencing and practicing them. This is how we develop our morality. As these are natural phenomena, we have no need to look for supernatural explanations beyond nature to learn both moral and immoral behaviors and activities. Information provided by scientific method can explain the natural phenomena that are superior to the religious explanations.

Each religion has its own peculiarities, but all religions lack universal applicability and acceptance, as they are unscientific and culturally dependent. Humanism has the ability to address the questions of life, the world, and nature in terms of realistic approach, which is common to everyone in every society. The ethical principles of humanism are shared values, unlike closed, religion-based values. Humanism is evergreen, always open to improvement as knowledge evolves and progresses in human society. Its secular entities go beyond religion and reach the global community with a common tenet rather than separate and competitive manner by various religious beliefs and faiths; each proclaims its own ritualistic merits. This is the beauty of humanism, and this is why it is better than religion.

Humanism unites all people under one umbrella called humanity into one holistic community; it inspires global ideals and standards. As a naturalistic philosophy, it recognizes the fact that life on this planet is the only life (as far as we till now) and that its diversity is precious. In this regard, humanism considers humankind as just one species among many with no right to do harm to anyone and any species and the environment on which we all depend. Unlike religion, humanism rejects the idea that this life is just a testing ground or a transit platform on the journey for a better life after death. On the contrary, humanists consider this life as the only one life we have. They further emphasize that it is our world and that everything in it is final for us and that it is, therefore, our responsibility and obligation to pass on this planet to future generations in the best possible condition we can. There had been, and are still today, clashes and hostility concerning the reality, the truth, and what we mean these to human life. Clashes occur not only between religions and nonreligious groups but also within religions. It is essential, in the interest of global humanity and for the future of our global society that we find some common and unique values aimed at living peacefully. To resolve this, humanism proposes that we should put aside religious differences, concentrate on our similarities, and strive for humanity. The model I demonstrated in the earlier section may be a good one.

I would agree with what Pat Duffy Hutcheon concluded in her book *The Road to Reason*. She said that the great idea and the source of power to survive are the premise of commonality and continuity among all existing inorganic and organic forms, which assert that humans are part of the universal process and no less natural than any other part. It requires no presumed injection of an unknowable spirit component at any point in the process of emergence. It offers no mysterious access to a consciousness beyond that created by our experience of nature. Rather, it implies that human actions and relationships are as subject to causation as those of any other existing entities, and that is the philosophical premise of evolutionary naturalism.

Good Life without Religion

Let me start with these words: "I appeal as a human being to human beings. Remember your humanity, and forget the rest," Bertrand Russell said. His message is simple—humanity is all and all about good life. One does not need anything but humanity to have a good life. We make our own meaning and purpose for our own existence. We choose to be kind and compassionate by ourselves. We know our responsibility to make the world a better place for all. We have science, technology, knowledge, intellect, and reason to help guide us. We decide right and wrong based on real-world experience but not on tradition or on falsehood reward or on punishment from a god.

But there is an irony that religious believers hold unjustified prejudice about nonbelievers. We, as humanists, want people to know what being nonreligious really means and, thus, want to change the unjustified prejudice that believers feel toward nonbelievers. The myth that a life without belief in a particular religion leads to loneliness or a lost sense of purpose or meaning can be disproved. There are about 1.5 billion people who do not associate with any religion who are humanists. This huge number of secular humanists is living a good life without religion.

We all want to live meaningful lives that are worthwhile, productive, and fulfilling. But what makes our lives meaningful and fulfilling? While some will say religion is the answer, others will disagree and don't find any meaning in religion because they can't believe in it. So it is not the religion but perception of life that matters. The fact is lives are not doomed to meaninglessness for those who do not believe in religion. Meaning in life without religion can be found in the many ways that do not reflect on religious faith or elements that are not religious in nature, such as friends, family, romantic love, good careers, learning, teaching, noble causes, caring for and helping others, striving for moral excellence, notable achievements and experiences, good hobbies and recreational activities, and so on. There are so many nonreligious sources of meaning in life, and by the virtue of those things, life can be meaningful without religion. For many, religion gives a sense of purpose of life, but it doesn't have a monopoly in giving a sense of purpose in life; rather, we can give a good purpose to our own lives.

People have found meaning of life and become moral even without religious faith and observance. Many people are bothered by the thought that without religion or a belief in a higher power to punish us, we would be disgraceful and awful. But humanism teaches that we all have the consciousness in us when we are born; people are basically good. Yet religion teaches us that belief in a higher power and religious observance are necessary for morality. This idea of religion toward morality is flawed. Many religious people do terrible things. They are sinners and are poor examples of religion and argued that God's grace will save them in the end because they are the believers. It sounds like that their belief has not given them a moral path.

Can we live better without religion? Yes, we can. We can live better without learning how to justify persecution of those who are different. We can live better without the divisions created by religions. We can live better without the attempts of religions to take away human rights. We can live better without the limitations religions pose to science, new thought, questioning of religion per se, the unknown, and personal exploration. We can live better without the prejudice and racial discrimination that religions propagate even to this day. We can live better by thinking that there is no heaven or hell other than the issues and matters we face on our journeys through life and the energies we generate to respond to them.

For many, when they are asked what you believe, the answer would be "myself." If I get into trouble, there's no God or Allah to help me out. I have to do it myself. We don't need to believe in an invisible sky god to find happiness; it has been inside us all the time. Religion tells us to put down our heads in the so-called house of God—that is, the church, mosque, temple, synagogue, etc. Going to these places and asking God for our salvation or finding the purpose of life is not a necessity; we can do so by doing good work in our day-to-day lives. The key issue is that it is not the people who are despised but that it is religious belief that is being despised. Dalai Lama said, "It becomes clear that no single religion can satisfy all humanity, so it is better to serve all humanity without appealing to religious faith." It echoes the idea that life is good without religion. We don't need divine commandment to know right or wrong. Our basic moral rules are common to all human beings. We can choose and set goals for our lives and be kind, compassionate, caring, and loving. These virtues have intrinsic values. We recognize our responsibilities at the individual and collective levels for making the world a better place for everyone without religion. We have science and reason, self-conscious and rational thinking to help guide us. Practically speaking, in everyday life, we decide right and wrong, what to do and what not to do based on real-world experience. Just think; about one-fifth of the world population, who are secular humanists, are living good lives without religion. More and more people are coming out as nonreligious all the time. In the past, people used to think that not having a religion meant being alone, but that's just not the case anymore. There are secular humanists or free thought groups near you that have regular meetings, activities, and family events.

Not only is a good life without religion possible at the individual level, but it is also possible at the society level. Phil Zuckerman, professor of sociology and secular studies, blogged the article "Imagine No Religion: Can a Society Be Successful without It?" In the article, he cited Denmark, where religion is very weak and marginal. This is a country where almost everyone accepts the evidence supporting evolution and believes that the Bible was written by humans and not the divine and where nearly everyone understands that morals and values exist independently of religion. His personal experience as a sociologist suggests that from economic well-being to educational attainment, from low violent crime rates to low unemployment rates, from quality of life to happiness and life satisfaction, this country is among the best on earth based on every standard measure and every international index. People there find good meaning in their lives without religion. Men and women find meaning in their relationships with family and friends, in their work, in their love of the outdoors, or in their hobbies and personal pursuits. Religion is simply not the only thing on this planet that can provide people with a profound sense of purpose, said Zuckerman. He concluded by saying that people can be honest, respectful, and decent without religious faith. Denmark is not only one of the safest places on earth, but it is also one of the most moral and ethically conscious cultures in the modern world.[131]

Creating Change through Humanism

The change through humanism is obvious. The first and foremost reason is that religions are based upon divine revelations of ancient people and old scriptures written by people, although it is said that the words of scriptures are the words of God. That millennium-old texts are bound to be antiquated, compared to the humanists' perspective of the nature of reality, which is progress for humanity based on the best available evidence and reasoning, as Roy Speckhardt, author of the book *Can We Create Change through Humanism*, has said. The basic questions the humanists and the religious people ask are based on two different tenets. As he has described, the questions that humanists ask are "Do you explore science, literature, and art with an open mind?" "Do you value your own experience and worth?" and "Do you act for the good of humanity, including future generations?" The questions that the religious people ask, as opposed to the above, are "Do you give exclusive allegiance to a unique prophet or savior?" and "Do you believe in an absolute, a personal god, an immortal soul?" One can compare the sets of questions and find answers according to their perspectives. Which ones sound reasonable to you in practical sense? If the questions asked by the humanists are good for people, then surely humanism can make the change into people's lives. On the contrary, if the believers' questions hold, then science will not progress, no human development will take place, and the human civilization will reach stagnation. But this is not going to happen because the natural world is dynamic and so is the human. As nature goes through evolutionary process, so do we. Think awhile where we were a millennium ago and where we are today in the twenty-first century. This does not need any explanation.

It is understood that talking about what we believe or don't believe has no solid resolution in any forum or table. What is noticed is that the nonbeliever arguments are taken negatively and that it is always difficult to talk about the nonbelievers. Therefore, it is a better option to find the appropriate word that does bear negative connotation but as an alternative. This alternative is humanism. This is why Roy Speckhardt has pointed out that the term *atheist* is receding from popularity as being a negative term that denies religious belief, but, instead, the term *humanist* has positive associations, and it is gaining popularity at a high rate. So our effort to help humanity is very much meaningful to the extent that it can bring change in our lives. So let us use the term *humanity*, as opposed to all other terminologies used for nonbelievers.

Who will not agree with the fact that social change takes place when innovative thinkers bring new ideas that eventually gain widespread acceptance? How? Social change occurs whenever any good new ideas and inventions change the way that people behave and interact and when new discoveries change the way of life and when people see their place in the ever-changing world and, thus, find meaning of life in that world and the only world they have. When discoveries, inventions, and social and economic progress were generally welcomed by social reformers, intellectuals, and distinguished people, social and religious conservatives never wanted or trusted those social changes. They thought that any change to the traditional way of living would demonize their belief and might even lead to a bad situation, as those changes were not revealed by God. One may argue that their real motive for resisting change was that they feared losing their privileged positions of power and wealth or that they truly feared moving away from religious belief. Whatever it is, they rejected the idea of change that was brought in the society by the social reformers whom we call humanists.

In the past, scientific development and technological advancement were very slow, and new discoveries were rare. Cultures remained the same or changed so slowly that the change was barely noticeable to all. So without having witnessed any significant change or even not being imagined as to how change might happen, people had the impression that the world would remain more or less the same forever. Having this kind of situation, any hope for change rested upon the divine intervention. The term "God knows" was being coined in the minds of the people in the past. Unfortunately, though, religious believers still think the same. This is an easy submittal to ignorance and stupidity.

We are optimistic in our thinking and action and, thus, think that there is change by humanism. The advancement of scientific and rapid technological and economic progress is unstoppable. At the same time, people are moving away from religion, especially the younger generation, and leading good lives without religion. This is the perfect recipe for change, not the blind faith but accepting reason. Those who are in this process of change think that there is nothing to gain by promoting religious blind faith and myths over scientific knowledge and technological progress. They think that life is more valuable and meaningful by relying on the best of scientific and moral knowledge, which unites (instead of divides) the people behind a common sense of purpose, the advancement of modern civilization and human progress for greater good. We can bring change into our lives through building humanism as a way of life that promotes progress by reason. What we witness today is that as our knowledge and experience grow, our technology continues to improve. As our understanding is getting more and more refined, a common worldview will emerge, which every clear and open-minded rational thinker can easily agree with. This common worldview is humanity. Another reason for creating change through humanism is that people want freedom from religious moral oppression and orthodoxy. People are not willing to accept metaphysical considerations of religious dogmas that are founded with no evidence, rational justification, and reason; rather, they are lies and stained by blood.

Humanism is not an ism in the sense of unquestionable doctrine like communism, socialism or Christianity, Islam, Hinduism, Judaism, etc. There is no such thing like "convert" to humanism. Instead, humanism is a certain range of beliefs, principles, and values. Whether or not you share those values and attitudes, you more or less become a humanist. Taken together, humanism is a set of beliefs, values, and ways of meaningful, good life that constitute a view of the world—a philosophy by which people live their lives. This is so simple, and because of its simplicity, humanism can bring change in our lives. Going to the church, mosque, synagogue, or temple will not serve humanity. Rather, caring about the suffering of others who are sharing this world with us will be the service to humanity.

The famous Bengali poet Kazi Nazrul Islam wrote about equality and humanity. In his poem "Of Equality I Sing," he wrote:

> I sing the Doctrine of Equality
> Which levels up all barriers and distinctions
> Places on the same of footing the Hindus-Buddhists-
> Muslim-Christian,
> I sing of the gospel of equality
> Who are thou? A Parsi? A Jain?
> A Jew? A Santal or Bhil or Garo?
> A Confucian? A disciple of Charbak?
> Say, if anything else!

In his poem "Human Being," we find him even more forceful in his words about human rights and discrimination. He wrote:

I sing of equality . . .
Caste, creed, religion—there is no difference
Throughout all ages, all places,
We are all a manifestation
Of our common humanity.

This is about the cry on behalf of the distressed global humanity. This is about not understanding why the humans once identified with the name of prostitute, peasant, laborer, beggar, woman, untouchable, communist, or any other titles should be less morally noble or less socially entitled or should be undermined, rejected, and even killed. He also thought about the good way of thinking in the context of humanity. He wrote in his poem "Coolies and Labourers":

"Let the heaven dawn upon all; let the people of all walks of life and countries stand on the shore and listen the flute-call of unity; and if one human being is hurt, then let the whole of humanity feel the pain equally. If one human being is humiliated, then let it be considered humiliation for the whole of humanity" (Langley, 2009). We see the spirit of the poet that any threat to any human being in terms of starvation, violence, homelessness, or social ignorance is a threat to humanity.

Humanists are concerned for the well-being of all, are committed to diversity, and respect those of differing yet humane views. We work to uphold the equal opportunity and enjoyment of human rights and civil liberties in an open secular society and maintain that it is a civic duty to participate in the democratic process and a planetary duty to protect nature's integrity, diversity, and beauty in a secure, sustainable manner. Thus, engaged in the flow of life, we aspire to a humanist vision with the informed conviction that humanity has the ability to progress toward its highest ideals. The responsibility for our lives and the kind of world in which we live are ours and ours alone.[132]

Given all the ingredients of humanity (as discussed all the way and as the key thesis of the book), we make the case. Yes, we can make a difference in our lives through following the principle, rich heritage, good moral, and values of humanism. I urge you to move up, look around, and make full use of humanism; it will change your life, a good, productive life with meaning.

Chapter Three Conclusion

Humankind has evolved as a social and cooperative species. They live and work together. As they depend on one another, it is important to treat other people as they would like to be treated themselves and to work together to solve their problems. Humanists ask themselves the same questions as everyone else, such as why we are born, what the purpose of life is, how life began, why we die, and what will happen after death. Religious people answer based on faith in God, but humanists look for answers based on reason, experience, and shared human values. Humanists look for evidence before they believe things, so they are likely to believe what scientists or their own experiences tell them or to remain open-minded about questions rather than to believe what someone else says. Humanists think about these questions for themselves. Well, some questions may not have answers, or the answers may not be looked for. In any case, the answers based on religious doctrines to those questions may not be the best ones.

Humanists believe that this is the only life of which we have certain knowledge and that we owe it to ourselves and others to make it the best life possible for ourselves and all with whom we share. People are free to think for themselves, to use reason and knowledge as their tools, and they are best able to solve this world's problems without a supernatural, powerful sky god or gods. Humanism is an appreciation of the art, literature, music, culture, knowledge, intellect, and creativity, which, if nourished, can continuously enrich our lives. Humanists always take responsibility for their own lives and take full credit of being part of new discoveries, seeking new knowledge, exploring new options. Instead of finding solace in prefabricated answers to the great questions of life, humanists enjoy the open-endedness of a quest and the freedom of discovery that this entails, as stated by the *Humanist Society of Western New York*.[133]

Humanists value happiness, freedom, and justice. They are motivated by the desire to increase these attributes aimed at making the world a better place to live for mankind. On situations that suggest not moral or not ethical, humanists look for evidence, the consequences of the action, and the rights of those involved, trying to find the kindest but rational course of action. Sometimes, humanist perspectives on moral and ethical issues are not very different, though, from those of liberal-minded religious people. However, a humanist view is explicitly based on reason, experience, empathy, and respect for others rather than on the texts of holy books.

Shared human nature is an important element in the context of developing consensus on the areas of religions, societies, and ethical and legal systems, about what is good or bad, tolerable or intolerable, moral or immoral, even when they disagree about where their values came from. A good example is the Universal Declaration of Human Rights (a declaration of shared human needs and values), which has gained wide acceptance and support internationally. Humanists believe in the "Golden Rules," which is "Do not treat others as you would not like to be treated yourself."[134]

The Golden Rule is not one philosophy to be claimed. It evolved in many communities in the past, which was used to resolved conflicts. Throughout the ages, many thinkers and spiritual traditions have promoted the Golden Rule in different versions of it. Here are some examples:[135]

- Do not to your neighbor what you would take ill from him (Pittacus 650 BCE).
- Do not unto another that you would not have him do unto you. Thou needest this law alone. It is the foundation of all the rest (Confucius 500 BCE).
- Avoid doing what you would blame others for doing (Thales 464 BCE).
- What you wish your neighbors to be to you, such be also to them (Sextus the Pythagorean 406 BCE).
- We should conduct ourselves toward others as we would have them act toward us (Aristotle 384 BCE).
- Cherish reciprocal benevolence, which will make you as anxious for another's welfare as your own (Aristippus of Cyrene 365 BCE).
- Act toward others as you desire them to act toward you (Isocrates 338 BCE).
- This is the sum of duty: do naught unto others that would cause you pain if done to you (Mahabharata [5:1517] 300 BCE).
- What is hateful to you, do not to your fellow men. That is the entire Law; all the rest is commentary (Rabbi Hillel 50 BCE).
- Thou shalt love thy neighbor as thyself (Lev. 19:18 1440 BCE).
- Therefore, all things whatsoever ye would that men should do to you, do ye even so to them (Jesus of Nazareth circa 30 CE).

Whichever ways one can express, humanists have taken the universal nature of this rule for granted in practicing this rule. What is important to remember is that humanism is a nonreligious ethical ideology. It is a way of life and a way of thinking to live in this complex world with adherence to strong ethics and moral values and emphasis on human rights. Humanists work responsibly with dedication, commitment, and compassion for a vision to make a better world for mankind and for other creatures as well.

Humanism may be classified in various ways based on different philosophies and propositions driven by human preferences, interests, values, and ideologies. Corliss Lamont, in the book *The Philosophy of Humanism*, defined humanism and classified it in ten different ways. They are as follows:

- Humanism rests on the idea of naturalistic metaphysics or attitude toward the universe that considers all forms of the supernatural as myth but, rather, regards nature as the totality of being and as a constantly changing system of matter and energy, which exists independently of any mind or consciousness.
- Humanism draws upon the laws and facts of science. Humanists believe that (a) human beings are an evolutionary product of the nature of which we are a part; that (b) the mind is indivisibly conjoined with the functioning of the brain; and that (3) as an inseparable unity of body and personality, humans have no conscious survival after death.
- Human beings possess the power or potentiality in solving their own problems through reason and scientific method.
- In opposition to all theories of universal determinism, fatalism, or predestination, humanists believe that human beings possess genuine freedom of creative choice and action and shape their own destiny.
- Humanists believe in an ethics or morality that grounds all human values in this earthly experiences and relationships and that holds as its highest goal this worldly happiness, freedom, and progress (economic, cultural, and ethical) of all humankind, irrespective of nation, race, color, or religion.
- Humanists believe that the individual can attain the good life through good and hard work and can achieve personal satisfactions and make continuous self-development, and at the same time, they can contribute to the welfare of the community by work, commitment, compassion, and love.
- Humanists believe in the development of art and beauty of nature's loveliness and splendor so that the aesthetic experience may become a pervasive reality in the lives of all people.

- Humanism believes in a far-reaching social program that stands for the establishment of democracy, peace, full freedom of expression and civil liberties, justice, rule of law, and a high standard of living on the foundations of a flourishing worldwide economic order that is good and just for all.
- Humanism believes in the unending questioning of any aspects of human life, including religious doctrines and dogmas.
- Humanism is not a new dogma; rather, it is an evergreen philosophy, newly discovered facts, and more rigorous reasoning.

We can name these humanistic philosophies as scientific, secular, naturalistic, democratic, or whatever. By any name we may call, humanism is a worldview that tells that we have only one life and that we should make the most of it within our ability and power in terms of creative work and happiness. To achieve this, we use our own knowledge, intelligence, reason, and justification without any sanction or support from supernatural sources and sky gods. Human beings can establish an enduring love, peace, beauty, and harmony among us on earth, so each one of us is responsible for human affairs, other beings, and the resources of our common shared planet. Our vision for a good world is one in which every individual's worth and dignity are respected, nurtured, and supported.

Humanist Institution deans[136] said that whichever humanist group among various types we belong, all of us are deeply concerned by these human attributes and that the challenges we are faced with cannot be passed on to any "higher powers." Deities and gods are human creations giving the idea that those higher powers would help believers feel secure in this ambiguous world. However, it has been seen that the beliefs in higher powers have often been the source of tribal, national, and global conflicts and violence. As we are the bearers of the historic Enlightenment, we have to continue to cherish values of freedom, reason, and tolerance, and it is our responsibility to develop this heritage for future generations.

Our planetary community is facing serious problems that can only be solved by cooperative global action. In a rapidly changing world, fresh thinking is required to move civilization forward. Humanity needs to reconstruct human values in the light of scientific knowledge. We are concerned with reconstructing old habits and attitudes to make happiness and well-being available for every person interested in realizing the good life for self and others. This is what neo-humanism is, a daring new approach, said Paul

Kurtz.[137] He presented a set of statements from *Neo-Humanist Statement of Secular Principles and Values*. They are as follows:

The neo-humanists

- aspire to be more inclusive by appealing to both nonreligious and religious humanists and to religious believers who share common goals;
- are skeptical of traditional theism;
- are best defined by what they are for, not what they are against;
- wish to use critical thinking, evidence, and reason to evaluate claims to knowledge;
- apply similar considerations to ethics and values;
- are committed to a key set of values—happiness, creative actualization, reason in harmony with emotion, quality, and excellence;
- emphasize moral growth (particularly for children), empathy, and responsibility;
- advocate the right to privacy;
- support the democratic way of life, tolerance, and fairness;
- recognize the importance of personal morality, goodwill, and a positive attitude toward life;
- accept responsibility for the well-being of society, guaranteeing various rights, including those of women, racial, ethnic, and sexual minorities, and supporting education, health care, gainful employment, and other social benefits;
- support a green economy;
- advocate population restraint, environmental protection, and the protection of other species;
- recognize the need for neo-humanists to engage actively in politics;
- take progressive positions on the economy; and
- hold that humanity needs to move beyond egocentric individualism and chauvinistic nationalism to develop transnational planetary institutions to cope with global problems. Such efforts include a strengthened world court, an eventual world parliament, and a planetary environmental monitoring agency that would set standards for controlling global warming and ecology.

We may ask a simple question: is humanism important to the future of the human species? Let us have a closer look into it. Humanism advocates secularism, which is the basis of modern sociopolitical structure, and suggests that religion should not be allowed to be authoritarian and must not dictate how societies should operate. Modern civilization is founded on secular humanism, which promotes freedom of speech, democratic values, civil rights, equality, justice, scientific inquiries, knowledge, intelligence, and so on. But the forces of conservatism and religious dogmas have always attempted to block these values to flourish, since most of the world religions are based on scriptures so old that they have no relevance to the contemporary world conditions and they are unable to provide any solution in response to social and natural problems. Moreover, fundamentalists adherent to those texts in the scriptures use them to justify war, violence, and social and ethnic unrest. It is of utmost importance that we have to change the way we did things in the past and adopt new things that modern science and technology offers and make conscious decisions ourselves rather than believe and practice the Dark Age religious dogmas.

The lesson we learn from the great thinkers and notable humanists worldwide whose principles and ideas shed light on what humanism is. A common theme appears to be the premise of evolutionary naturalism, as opposed to spiritualism, transcendentalism, or any other form of dualism. Pat Duffy Hutcheon noted that the teachings of those great thinkers give three additional thoughts: first, "human know-how" and the universality of scientific approach as a means to build knowledge; second, human imagination and creativity and technical skill; and, third, overriding focus on morality as the unique responsibility of mankind. This is the most likely modern worldview of humanism that would save the new generation of population in the global village from self-destruction at present and in the future (Hutcheon 2001).

Humanism is an alternative life stance for those who choose to live without religion. It's an ancient concept with origins dating back to the ancient Greeks, Confucius in China, and Vedic philosophy in India. Humanism links to free thought, ethics, rationalism, and secularism. Realistically, all we have is this life and this world, and accepting this fact enhances our lives. Religions have not served the world well other than giving personal consolation. When question is raised by religious defenders: what could we think of our life stance in terms of higher morality, ethics, and values in the world we live in than religion? The obvious answer would be something far better, deeper, kinder, warmer, loving, passionate, and far more rational called humanism. As we discussed earlier, humanism is the universal religion. Seven billion people on earth do not belong to one religion, but all believe in humanity. Would it not be better to join humanism? It also resonates Bertrand Russell's famous quote "Remember your humanity, and forget the rest." I think we can forget the rest and embrace humanism. Moreover, the humanism has been best expressed by the philosopher Baruch Spinoza when he said that peace and tranquility among human is not the absence of war, conflict, or violence; rather, it is a virtue, a state of mind, a disposition for benevolence, confidence, and justice for all. This vision, perhaps, is a universal compass that we must tailor to the realities of our times.

CHAPTER FOUR

Transhumanism

We are at the very beginning of time for the human race. It is not unreasonable that we grapple with problems. But there are tens of thousands of years in the future. Our responsibility is to do what we can, learn what we can, improve the solutions, and pass them on.

—Richard Feynman

The human race is just getting started . . . The cerebral cortex is only a hundred thousand years old. It's still a baby, sucking teat and eating Cheerios. We might get better, maybe even wise, if we can last another thousand years

—Ellen Gilchrist

Meaning, Theory, and Values

An Overview

The human species is still very young on planet Earth, and we have yet to see a lot of what is possible to know and to become in the foreseeable future. But success in this enterprise is far from assured, because we still have our limited human wisdom, knowledge, and compassion to guide us through the transition. Our efforts with the use of development of human-enhancement tools, to eliminate or minimize the major catastrophic risks, and the works to alleviate the sources of human suffering would be our job to improve the human condition. Our future knowledge should be more radical so that it can promote not only traditional means of improving human nature, such as education and cultural refinement, but also direct application of modern medicine and more advanced technology to overcome some of our basic biological limits. In fact, this is happening at present, when we closely look into the advancement of science and technology. Technology is gradually making space over biological processes. Although many of the technologies can be foreseen, we do not know yet how long it will take to fully develop them. However, we can foresee the future that humans (as we are today) are not a finished product. Rather, they are evolving organisms, waiting for the right conditions to blossom. It seems that humans can evolve beyond natural and biological limits.

Nick Bostrom (2005) mentioned in an article that the human desire to learn and acquire new capacities is as ancient as our species itself. Humans have always sought to explore the unknown and to expand the boundaries of our existence, whether it is social, geographic, or mental. Examining Darwin's *Origin of Species*, we can argue that the current version of humanity is not the end point of evolution but, rather, just an early phase. The rise of physical science also has contributed to the foundations of the idea that technology could be used to improve the human organism, he said. The effort to understand and find ways to reduce existential risks has been a key concern for some transhumanists such as Eric Drexler and Eliezer Yudkowsky. Nick Bostrom pointed out that both bioconservatives and transhumanists agree that technology can be used to substantially transform the human condition even in the twenty-first century. However, both camps agree that this raises the question of obligation about practical and ethical implications, and they are concerned about the medical risks of side effects. Bioconservatives draw attention to the possibility that subtle human values could get eroded by technological advances, and they warn that transhumanists should be sensitive to these concerns. However, transhumanists emphasize the enormous potential for genuine improvements in human well-being and human flourishing that are attainable only via technological transformation. Transhumanists advocate that bioconservatives should appreciate the fact that humans can realize great values by venturing beyond our current biological limitations (ibid.).

With no doubt, nature is creating countless events that can be considered criminal acts against humanity, such as plagues and diseases, earthquakes and floods, pests, poisonous plants, aging, and so many. Nature has created us to suffer and die. But we have been and will continue to fight against nature's crimes and adopt its precious things by using human intelligence and man-made technology. When we will win the battle against those crimes of nature, we will be more than humans, and we will be better than humans as we are today. Our top priority now is to fight aging and all diseases. When we can achieve such a noble and beneficial goal, we will have a world of opportunities where we can greatly improve and upgrade our lives using advanced technologies such as genetics, cybernetics, nanotechnology, etc. We will get rid of the anthropomorphic values and ideals that limit our imagination, and we will become posthuman beings eventually. We seek to evolve ourselves. We, as transhumanists, will not impose anything on others like theologians do. Rather, we will strongly emphasize to use technologies that will lead to the enhancement of the human race. We know that perfection does not exist in this world; neither can we be perfect as human beings. But we can be better. Evolution of human species is no longer a natural process. They are expected to take their evolution as their own task. For modern human beings, the "sky is the limit"; we can conquer the sky.[138]

Understanding Transhumanism

It is not right to think that current humanity is the end point of evolution. So the term *transhumanism* comes into being. Transhumanism is an international intellectual and cultural movement supporting the use of new sciences and technologies to enhance human cognitive and physical abilities, which, in turn, find answers to the vital sufferings from undesirable and unnecessary human condition, such as disease, aging, and death. As mentioned in the introduction chapter, it is a new philosophy that promotes the ideas of humanism in a new world where science and technology are the driving forces of change. Julian Huxley, a biologist and humanist, the first director general of UNESCO, and founder of the World Wildlife Fund, is the first person to use the actual word *transhumanism*. He wrote the following:

The human species can, if it wishes, transcend itself—not just sporadically, an individual here in one way, an individual there in another way, but in its entirety, as humanity. We need a name for this new belief. Perhaps transhumanism will serve: man remaining man, but transcending himself, by realizing new possibilities of and for his human nature.

He published his essay "Religion without Revelation" in 1927, which was later reprinted in his book *New Bottles for New Wine*, published in

1957. Huxley has defined transhumanism as "man remaining man, but transcending himself, by realizing new possibilities of and for his human nature." Other scientists and philosophers discussed similar ideas in the first half of the twentieth century, and those ideas gradually helped create new philosophical movements that considered nature and humanity in a continuous state of flux and evolution. English scientist John Burdon Sanderson Haldane and French philosopher Pierre Teilhard de Chardin also helped identify new trends in the future evolution of humanity.[139]

Transhumanism is a cultural and intellectual movement that is likely to improve the human condition through the use of advanced technologies. One of the core concepts in transhumanist thinking is life extension. Through genetic engineering, nanotech, cloning, and other emerging technologies, eternal life may soon be possible. Transhumanists are interested in the ever-increasing numbers of technologies that can boost our physical, intellectual, and psychological capabilities beyond what humans are naturally capable of at present. This is why the term *transhuman*. Transhumanism is a new thinking. Whether it is brain implants, bionic body parts, test tube babies, robotic hearts, exoskeleton suits, artificial intelligence, or gene therapies or genetic engineering that is likely to eliminate natural death, it's obviously an uncharted territory for the human species. Humans can no longer be regarded as a stable and static category as one that occupies a privileged position in relation to all that is included under the category of the nonhuman.

On the contrary, humans must be understood as a tenuous entity that is related to the animal, the "natural," and, indeed, other humans as well. Humans are at a crossroads like other natural species that are reclassified in the face of new dynamics and shifting epistemological paradigms. Moreover, such dynamics and interpolation serve to reveal the boundaries of humans as a corporal and cognitive construct. Discovering such boundaries, one may gather the knowledge of where humans end and where humans stand to expand themselves or become more than human. Our understanding about ourselves and about our relationships with nature around us has increased significantly because of the continuous advances in science and technology. Reality is not static, since humans and the rest of nature are dynamic, and both are changing constantly. Transhumanism transcends such static ideas of humanism as humans themselves evolve at an accelerating rate. In the beginning of the twenty-first century, it is now clear that humans are not the end of evolution but just the beginning of a conscious and technological evolution.

Since English naturalist Charles Darwin first published his ideas about evolution on *The Origin of Species* in 1859. It has become clear to the scientific community that species evolve according to interactions among them and with their environment. Species are not static entities but dynamic biological systems in constant evolution. Humans are not the end of evolution in any way but just the beginning of a better, conscious, and technological evolution. The human body is a good beginning, but we can certainly improve it, upgrade it, and transcend it. Biological evolution through natural selection might be ending, but technological evolution has taken place and is accelerating in a certain pace. Technology, which started to show dominance over biological processes some years ago, is finally overtaking biology as the science of life.

A logic theorist, Bart Kosko, said, "Biology is not destiny. It was never more than tendency. It was just nature's first quick and dirty way to compute with meat. Chips are destiny. Photo qubits might also come after standard silicon-based chips, but even that is only an intermediate means for augmented, intelligent life in the universe." *Homo sapiens* is the first species in our planet that is conscious of its own evolution and limitations, and humans will eventually transcend these constraints to become enhanced humans, transhumans, and posthumans. It might be a rapid process, like caterpillars becoming butterflies, as opposed to the slow evolutionary passage from apes to humans. Future intelligent life-forms might not even resemble human beings, and carbon-based organisms will mix with a plethora of other organisms. These posthumans will depend not only on carbon-based systems but also on silicon and other platforms that might be more convenient for different environments, like traveling in outer space (Cordeiro).

Transhumanists advocate the improvement of human capacities through advanced technology that helps eliminate disease, provides cheap but high-quality products to the world's poorest, improves quality of life and social interconnectedness, expansion of life expectancy, and so on. As modern science and technology have attained even greater control over the tiniest structure of matter (at the atomic and subatomic level), the technological goals have become increasingly ambitious. In fact, new technologies make us happy in a long-lasting way—the Internet is a prime example. But how plausible is transhumanism? We know that in the 1930s, many sensible people were so certain that human beings would never get to the moon. This was just one of many predictions that turned out to be incorrect. Ironically, in the early twenty-first century, people do not know what will be possible in the near and far future. As we learn from current technological advancement, transhumanism is a strongly held belief among many computer geeks, notably computing guru Ray Kurzweil (a believer in the "technological singularity," where technology evolves beyond humanity's current capacity to understand or anticipate it), and Sun Microsystems founder and Unix demigod Bill Joy (who believes the inevitable result of artificial intelligence research is the obsolescence of humanity).[140]

The general expectation of the transhumanists is that in the near future, greater manipulation of human nature will be possible because of the adoption of highly sophisticated techniques that are likely to be available, such as machine intelligence greater than that of contemporary humans, direct mind-computer interface, genetic engineering, nanotechnology, etc. In fact, certain recent technological advances tend to believe that the realization of transhumanist ideas appear more plausible. Scientists have developed an implant that can translate motor neuron signals into a form that a computer can use, thus opening the door for advanced prosthetics (artificial body parts) capable of being manipulated like biological limbs and producing sensory information. This is on top of the earlier development of cochlear implants, which translate sound waves into nerve signals, which are often called bionic ears.[141]

Transhumanists hope that with responsible use of science, technology, and other rational means, we shall eventually manage to become posthuman with greater capacities compared to the present human beings. Posthumanism considers the possibility that historical phenomena (such as advances in technology and science or discoveries about animals) are leading to fundamental changes in the human species and its relationship with the world. In fact, the term *posthuman* is used to describe the next stage of human development. In other words, we can use technology to transition ourselves to something beyond human. The fact is, transhumanism reminds us that the human body and society can be systematically altered with technology. On the other hand, posthumanism emphasizes the fact that while we use technology and consider its effect, we have to take into account the countless variables and voices that need to be heard. Those are from our everyday life, such as different ethnicities, genders, and classes, to the extreme elements, such as robots, artificial intelligence, animals, etc. Therefore, both transhumanism and posthumanism, when defined in assimilation, integration, and conjunction, can draw a broader picture of humans, culture, and society that allows us to see that we are all trying to answer similar questions.[142] Given the current trend of continued developments of advanced science and technologies, foreseeable occurrence of posthuman is now being widely discussed and invoked as the inevitable next stage that humans are facing.

Transhumanist Technology

An Overview

For the last several decades, we have been living in the world of technology, without which we cannot progress. In fact, our nature as human beings has fundamentally changed by the advancement of technology. We are likely to see that in the next half century or so, the relationship between human and machine would be redefined in a profound way. Think for a while about how we lead our everyday lives and what leads us to make our day. The simple answer is, we have been profoundly integrating technology into our daily lives. Our society has embraced technology by necessity. How we conduct business; do everyday work in the office and outside; organize social gatherings; do shopping; share intimate moments and fall in love; even wage war and revolutions; produce art, science, and literature; share ideas and experiences; and whatnot—all have been redefined by the use of technology. Information sharing is a notable example as to how it works. This human activity is performed by billions of people in the world via digital medium, which did not exist a generation ago. Although the machine code is virtual and intangible, the information conveyed is real and meaningful. Surprisingly, the information we share with others simultaneously around the globe travels at the speed of light. Technology has become the means for all these things that make us human, without which we are not human. Christopher Phillips said, "Welcome to the future. You are now living in an era of transhuman technology—an era we may call an evolutionary renaissance. And you are what we call a cyborg: part human, part machine."[143]

Transhumanists support the emergence and convergence of technologies including nanotechnology, biotechnology, information technology, and cognitive science (NBIC), as well as future technologies like simulated reality, artificial intelligence, superintelligence, mind uploading, chemical brain preservation, and cryonics. Humanists believe that humans can and should use these technologies to become more than human. They support the recognition and protection of cognitive liberty, morphological freedom, and protective liberty as civil liberties to guarantee individuals the choice of using human-enhancement technologies on themselves and their children. There is also speculation that human-enhancement techniques and other emerging technologies may facilitate more radical human enhancement no later than at the midpoint of the twenty-first century. For example, Kurzweil's book *The Singularity is Near* and Michio Kaku's book *Physics of the Future* outline various human-enhancement technologies that provide insights on how these technologies may impact the human race in the near future and one hundred years from now. In a report, the National Intelligence Council has described major trends on technological developments that might be expected in the next thirty years or so. The council has made some significant predictions such as (1) the end of US global dominance; (2) the rising power of individuals against states; (3) a growing middle class that will increasingly challenge governments; and (4) ongoing shortages in water, food, and energy. At the same time, they also envision a future in which humans will significantly be modified by the use of science and technologies, which will herald the dawn of the transhuman era.[144]

Some reports on the converging technologies and NBIC concepts have criticized their transhumanist orientation and science fiction character. But technology continues to advance. For example, research on brain- and body-alteration technologies has been accelerated under the sponsorship of the US Department of Defense. The department is interested in the battlefield advantages that they could provide to the super soldiers of the United States and its allies. There has already been a brain research program to "extend the ability to manage information," while military scientists are now looking at stretching the human capacity for combat to a maximum of 168 hours without sleep. Neuroscientist Anders Sandberg has been practicing on the method of scanning ultrathin sections of the brain. This method is being used to help better understand the architecture of the brain. As of now, this method is currently being used on mice. This is the first step toward uploading contents of the human brain, including memories and emotions, onto a computer.[145]

In the article "Transhumanism and the Meaning of Life," Anders Sandberg concluded that although transhumanism does not have a unified theory of the meaning of life, certain themes that are linked to the different strands repeat again and again. He further said that there are objective values or goals that can make transhuman life meaningful and that there is a great deal of individual subjective choice in setting goals and determining how to reach them. Objectives are reducing suffering and unnecessary limitations and achieving well-being, wisdom, life, diversity, and an open future, which is the most common approach within individual transhumanism. In terms of posthumanity, our lives will be meaningful in the physical sense. In the infinite future, our lives will be much larger than the normal future, which also makes it meaningful. The transhumanists, no matter they are secular or theistic, contemplate within a meaningful worldview because of its scope. The key idea is that transhumanism links our current microscopic view with the macroscopic view—that is, the grandness of the universe unveiled by modern science and advanced technology. As the universe is so vast, the transhumanists would like to experience benefits and meaning as dynamic rather than static and increasing rather than decreasing.[146]

Let us take a look at the top-ten advanced technologies that can bring significant changes and transformations on human lives.

Topmost Technological Endeavor

The Cryonics

Cryonics is the high-fidelity preservation of the human body (particularly the brain) after death, which would be possible for future revival. Cryonics is a transhumanist technology that is not only already available today but also relatively mature. This technology can reliably stop cells from decaying. In vitrification, the brain is frozen with a cryoprotectant (antifreeze) mixture instead of a conventional manner, which effectively prevents the formation of crystals, causing the water to freeze smoothly like glass. Maintenance of a cryo-patient is not difficult, and it requires no electricity but merely the replenishment of liquid nitrogen every three weeks. As cryonics becomes more popular, this process could become automated and extremely reliable. For example, the Cryonics Institute in Michigan has operated since 1976 without a single mishap.

For the proof of existence of cryonic revival, there are frogs that can freeze solid and revive later. However, reviving a human from freezing would likely require molecular nanotechnology (MNT). Once we are able to develop MNT, the prospect of successful revival is extremely possible, and it would involve melting the ice and rebooting the metabolism by kick-starting the appropriate chemical reactions within cells.[147]

Virtual Reality

Virtual reality is a screenshot from the game Crysis. Looking at screenshots from the game, one can see that the computer graphics are already beginning to approach photo-realism. Sometime in the 2020s, reality simulations will become so high in resolution and immersive that they will start to get indistinguishable from the real thing.

Simulations will become the preferred environments for work and play. Pretty soon, the main obstacle to truly immersive VR will not be the visuals but the haptics, our sense of touch. To fool our senses into believing haptic technologies are conveying the real thing, the frame rate needs to be significantly higher for visual technologies, a few hundred updates per second rather than a few dozen—which is why development could take another decade or two. But many millions of dollars are currently going into efforts to develop advanced VR.

Clearly, World of Warcraft's eight million subscribers and Second Life's five million subscribers are onto something. These worlds typically outclass the real world in terms of customizability but still have yet to catch up in terms of sensory richness or social fulfillment. But it's only a matter of time. In the mid to late 2020s, one may expect full-body, high-quality, haptic VR suits to be affordable to the average person in developed countries, either obtained from your local store or perhaps printed right out of a desktop nanofactory. For more on this, one can read the scientific paper "Towards Full-Body Haptic Feedback."[148]

Gene Therapy / RNA Interference

Gene therapy replaces bad genes with good genes, and RNA interference can selectively knock out gene expression. Together, they give us an unprecedented ability to manipulate our own genetic code. By knocking out genes that code for certain metabolic proteins, scientists have been able to make mice stay slim no matter how much junk food they eat. Lou Gehrig's disease has been cured in mice. Although not developed yet, it may be not too far that the same therapy would be developed that can cure humans too. Aubrey de Grey's Strategies for Engineered Negligible Senescence (SENS) research program contains various prescriptions for the use of gene therapy. Within a couple of decades or so, progress in antiaging therapies will improve to the point where we will be gaining more than an extra year of life span per year, reaching "longevity escape velocity," eventually culminating in indefinite life spans.

Gene therapy is really exciting because it's just the beginning. No scientist has yet performed gene therapy on germ line cells (sexual cells in the gonads) because of the ethical controversy of producing genetic changes that are heritable, but it's only a matter of time. However, country regulations on gene therapy research and development will only be able to slow the overall progress of the field by a few years at most. In its mature form, gene therapy and genetic engineering will become extremely cheap and powerful, letting humans live comfortably in a wider range of environments and gain immunity to all diseases.[149]

Space Colonization

Space colonies will become necessary to house the many billions of individuals that will be born in the future as our population continues to expand exponentially. Marshall T. Savage, in *The Millennial Project*, has estimated that the asteroid belt could hold 7,500 trillion people, if thoroughly reshaped into O'Neill colonies. At a typical population growth rate for developed countries at 1 percent per annum (doubling every seventy-two years), it would take us 1,440 years to fill that space. Siphoning light gases off Jupiter and Saturn and fusing them into heavier elements for construction of further colonies seem plausible in the longer term.

The question is, why expand into space? The answers are blatantly obvious, but the easiest is that the alternatives are limiting the human freedom to reproduce or mass murder, both of which are morally unacceptable. Population growth is not inherently antithetical to a love of the environment. In fact, by expanding outward into the cosmos in all directions, we'll be able to seed every star system with every species of plant and animal. The genetic diversity of the embryonic home planet will be tiny by comparison. An important perspective is that space colonization is not only related to transhumanism through the mutual association of futurist philosophy but also more directly related, because the embrace of transhumanism will be necessary to colonize space. Human beings are not designed to live in space. Our physiological issues with space are manifold, from deteriorating muscle mass to uncontrollable flatulence. On the surface of Venus, we would melt; on the surface of Mars, we would freeze. Given that circumstance, the only reasonable solution is to upgrade our bodies. Not terra form the cosmos, but cosmos form ourselves.[150]

Cybernetics

The process of cyborgization has already been happening for centuries since the advent of clothing and piercings. For many generations, our technological gadgets have been getting smaller, more functional, and more closely integrated with our natural activity. Artificial limbs, ears, and organs are already available and continue to improve. This trend will continue in the future. Cyborgs already walk among us, and they look just like normal people. Many cyborg upgrades will become available in the twenties and thirties, such as hearing and vision enhancement, metabolic enhancement, and artificial bones and muscles and organs, and even brain-computer interfaces will be invisible to the casual observer, implanted beneath the skin. Cybernetic features on the surface, such as dermal enhancements or technological actuators like retractable wings, will be carefully camouflaged.

Recently, Microsoft announced Microsoft Surface, a mouse-less, keyboard-less form of desktop computing that takes input from finger tracing and hand gestures. The sophistication of biotechnology and the availability of better materials and precision manufacturing will let us make systems so small and effective that even everyday people will choose to implant them. These cybernetic systems will greatly improve our everyday experience, from letting us hear a wider range of ambient sounds to viewing millions of stars rather than just a few thousand to making us more resistant to accidents. They will improve the overall economy by enabling us to do more work in less time for better pay. In the long term, enhanced humans may get a bigger portion of the economic pie than unaugmented humans, but the pie itself will become so much larger that even the poorest humans of tomorrow will be better off than the wealthiest of today.[151]

Autonomous Self-Replicating Robotics

Why will you do manual labor when the robots can do it for you? Self-replication is the key to robotics. A landmark NASA study, "Advanced Automation for Space Missions," found that robotic self-replication is just a matter of engineering and that no fundamental theoretical breakthroughs are needed. The study proposed sending a hundred-ton package to the moon, with a self-replication time of one year, and letting it self-replicate until the desired level of development is attained. The design was based on electric carts running on rails within the factory, paving machines that direct sunlight to melt lunar surface materials (regolith), robotic strip miners for obtaining raw materials, and a solar cell canopy for powering it all. After ten years, over one hundred thousand tons of a lunar factory could be produced autonomously. The factory's functions could then be hijacked for the benefit of human colonists, used to produce housing and products and to provide large quantities of solar power.

If similar self-replicating systems could be constructed on Earth, it could provide plenty of material. Self-replicating factories could turn the vast, empty badlands of Australia into lush gardens by pumping water from the oceans, self-replicating factories in the high Arctic could melt snow and create transparent gigantic domes suitable for habitation, and submersible automatons in the seas could dredge sand from abiotic regions of the ocean floor and process it into gigantic platforms for human colonization. By opening up such vast new regions of the Earth's surface, overpopulation and crowding could be avoided for decades. Excess of population on earth can be moved to the moon, Mars, and the asteroid belt using the power of self-replicating robotics to create rotating space colonies suitable for housing trillions of people. Self-replicating factories could reduce the costs of material goods close to those of food. As such, humanity might actually shift from having a zero-sum perspective to a positive-sum perspective. With medical tools and basic goods in ample supply, no one in the world would need to suffer from poverty or curable disease. The nature of human work would shift from hard manual work and mind-numbing routine to more creative and personally fulfilling endeavors like art, music, math, science, literature, and exploration.[152]

Molecular Manufacturing

Molecular nanotechnology would use massive arrays of nanometer-scale actuators (produced initially through self-replication) to manufacture macroscale products with atomic precision. This concept is known as the nanofactory. The creation of nanofactories would mean that practically everything could be made out of diamond. The motors would become so powerful that a cubic centimeter would provide enough torque to propel a car, medical nanodevices could heal wounds and repair organs without the need for surgery, and air-suspended nanodevices (utility for) could be configured to simulate practically any desired object on demand.

Many of the prerequisites of molecular manufacturing have already been demonstrated. Molecular surgery has been used to snip off and replace individual hydrogen atoms, various functional nanoscale devices have been built, scanning tunneling microscopy has been used to mechanically manipulate individual atoms, and so on. The challenge is to create a nanoscale-manipulator arm capable of placing individual atoms with angstrom-level precision, avoiding undesired reactions, and serving as a universal constructor that can build a copy of it.[153]

Megascale Engineering

Most people are familiar with megascale engineering because it is seen throughout fiction—the Death Star, for instance. Typically, megascale engineering refers to building structures at least one thousand kilometers in length in one dimension, such as a space elevator, Globus Cassus, or Dyson sphere. With the self-replicating robotics described above, the production of such large structures could be done largely by autonomous drones, with intelligent agents only managing the highest, top-level functions and architecture. Megascale engineering is one of the transhumanists' magnificent visions—intelligent beings spreading across the cosmos and eventually shaping the very structure of the universe itself.[154]

Mind Uploading

Mind uploading (sometimes referred to as nonbiological intelligence) centers on the controversial proposition that cognitive processing can be implemented on substrates other than our current neurons. Considering decades of successful results in neurophysiology and the recent construction of the world's first brain prosthesis (an artificial copy of the hippocampus), this seems very likely. It appears that our minds are defined more by the information pattern they embody than by the particular hardware they are implemented on.

Mind uploading would involve simulating a human brain in a computer in enough detail that the "simulation" becomes, for all practical purposes, a perfect copy and experiences consciousness, just like protein-based human minds. If functionalism is true, like what many cognitive scientists and philosophers correctly believe, then all the features of human consciousness that we know and love—including all our memories, personality, and sexual quirks—would be preserved through the transition. By simultaneously disassembling the protein brain as the computer brain is constructed, only one implementation of the person in question would exist at any one time, eliminating any unnecessary confusion.

Many philosophers of mind have long recognized this, but the wider public has not accepted it; people don't want to think that they are just data structures being implemented as computational automatons on biological neurons. But it is hard to think of it any other way. Once we dismiss the possibility of an immaterial soul, we must acknowledge the mind as a material pattern implemented in physical configurations, and if other substances aside from our current neurons can meet the requirements for these configurations, then there is no reason why intelligence and consciousness could not exist on another substrate.

If our brains really don't have to be made out of meat, then we can transfer them to other substrates. By incrementally replacing each neuron with a synthetic neuron equivalent, the whole process could go down painlessly and seamlessly. The transfer could be as slow or as fast as we want—from the information-processing perspective of the brain itself, nothing ever changes. Light still comes in through the eye's lens, hits the retina, is transformed into nerve impulses, which travel down the optic nerve, and receives further processing in the visual cortex at the back of the brain, the highlights of which are sent to the prefrontal cortex for integration with information from the other senses. The brain can't tell if it's made out of traditional meat, accelerated biological neurons, or entirely nonbiological neuron equivalents—the computation is the same. This notion is also referred to as an application of the Church-Turing thesis.

If entirely synthetic brains are possible, then there's nothing that stops one from inhabiting computer networks, not indirectly, sitting in chairs as we currently do, but directly, engaging in computer worlds as a sentient program of tremendous complexity. With molecular manufacturing on hand, reversing the process would be as simple as printing out a hundred or so kilograms of flesh and bone again, complete with memories from the networked experience. If the process functions well, then virtual experience will be indistinguishable from physical experience. It would be even more enjoyable because of the newly accessible freedom. In a virtual world, there are no laws of physics except those we choose.[155]

Artificial General Intelligence (AGI)

Artificial intelligence such as thinking, feeling, imagining, creating, communicating, and thoughtful synthetic intelligences with conscious experiences are possible. Whatever computing is necessary is within the technological reach, and the present-day computing speeds are fast approaching the computing power of the human brain. The fastest present-day supercomputer, Blue Gene/P, is an example. It operates continuously at speeds of over a petaflop, which is a million billion operations per second.

Researchers are working diligently toward artificial general intelligence, informed by the mathematics of inference and probability theory—to name some, Jurgen Schmidhuber, whose main scientific ambition has been to build an optimal scientist; Marcus Hutter, author of the landmark book *Universal Artificial Intelligence*; Ben Goertzel, who presented his artificial intelligence (AI) design; and Eliezer Yudkowsky, who has developed a reflective decision theory. To be optimistic about the feasibility of general AI, these individuals will keep working and will eventually succeed.

If raw materials such as sand can be converted into computer chips and then into intelligent minds, eventually, lots of materials in the solar system could be made intelligent and conscious. The result would be a "noetic Renaissance," which is the expansion of intelligence and experience beyond our wildest dreams.[156]

Anyone who watched the science fiction movie *Elysium*, starring Matt Damon and Jodie Foster, understands how the separation between technology and humanity disappears. The movie showed that by installing computer chips directly into the human brain, it is possible to increase the capability for thought and to fix neurological disorders or create artificial limbs to allow human beings to exist without the limitations imposed by their natural physical form. Matthew Bulger reviews the film and its exploration of transhumanism. The key message is that human beings will eventually become their own master and will be able to remove the status quo of inequalities of society and human existence by eliminating physical and mental deficiencies by integrating advanced technology into the human body such as cyborgs. All human beings, regardless of race, gender, sexual orientation, religious beliefs, national identity, or any other factors, can benefit from the wonders of advanced technology and science—what transhumanism is meant. Thus, transhumanism serves the core values of humanity as it weighs equal value on all human life. Most importantly, transhumanism relies on human innovation by nurturing science rather than prayer or religious beliefs to uphold humanity.[157]

We have been in the evolutionary process whereby we become more and more integrated into our technology, and someday, we will reach a point where the lines between human and technology are going to diminish. It is obvious that those who identify as transhumanists are simply people who are looking forward to see this transformation sooner rather than later. As a matter of fact, we are already at a point where some people are alive because of technology, such as many people live with pacemaker, with artificial organs, or with blood transfusion, or take a drug that's keeping them alive and so on. Transhumanism represents the only logical, realistic thinking that could allow humans to extend their life spans and become posthuman, with greater capacities to reach the next stage of human development, which, in turn, replaces the need for blind faith and silly superstitions about afterlives, hell, and heaven. We see an exciting future where we will profoundly apply this technology responsibly to build a rational society in which every human being will have equal opportunity to fulfill their potential, irrespective of race, color, and ethnicity as well as physical and mental disposition.

Gregory Stock, author of the book *Redesigning Humans*, pointed out that although our technological advancement and developments of biology might be very complex to rework, the recent understanding of human genome and rapid progress to unravel genetic mystery suggest that modification of human biology is not too far from reality than the distant space travel we see in science fiction movies. But human genetic enhancement cannot happen without the help of computerized data analysis and communication technology. Artificial intelligence experts believe that computers will soon transcend humans. He further commented that we are probably the last humans in the sense that future humans will modify their biology to differ from present humans in a meaningful way. How will the future humans look back to our era? Stock has described a scenario. Future humans will see our time as primitive, when people lived only seventy to one hundred years and died of awful diseases. Children were conceived outside a laboratory by a random, unpredictable meeting of sperm and egg (that is, sperm and egg join together after sexual meeting of a man and a woman). Would it not be a privilege if we cannot only see but also participate in this unprecedented achievement?

I am convinced that the resultant force generated by the development of technology and biology in combination with humanity will help mankind become much better humans. Let us overview what scientific and technological development we are going to experience in the next century.

Science in the New Century

Predicting a century ahead is a difficult task. It is a challenge to dream about future discoveries and inventions through technological advancement that will alter our blind faith and beliefs and eventually establish humanity without religious boundary. Science is the engine that already started changing the foundations of civilization and leading us to a new century with wandering miracles. Only in the last few decades, so many scientific discoveries have been accumulated that have not been achieved in all human history. Scientists foresee that by the next one hundred years, the scientific knowledge and discoveries will add many times over (Kaku 2011).

We no longer live in the Dark Age. Today, we do not think lightning bolts, tsunamis, earthquakes, plagues, and diseases are the act of God or gods. Our ancient ancestors did not see the magnificent achievements of science and technology. For example, jet planes fly much higher up beyond clouds, rockets explore the moon and planets, MRI scanner peers inside the living body, cell phones keep us talking with anyone in the planet, and so many. Science is dynamic rather than static, and it is expanding exponentially. Scientific discoveries and innovations are changing in every aspect of human lives and social landscapes, which, in turn, is going to alter the old traditions, beliefs, and prejudices. Scientists believe that one hundred years from now, we will be able to manipulate objects with the power of our minds rather than by the power of God or gods (as orthodox people believe so). Here are some examples: Computers will be able to accomplish our wishes, and we will be able to move objects by our thoughts (a telekinetic power that belongs to God or gods). With the power of biotechnology, we will be able to create perfect bodies and extend our life spans. With the power of nanotechnology, we will be able to take an object and transform it into something else; we will be able to harness the unlimited energy of the stars and many others. While these are unimaginably advanced powers of mankind, which seem to be godlike power, the works of scientists have already been started in respective fields. The equations of the quantum theory are the basis of the creation of the universe, and the culmination of all upheavals is the formation of a planetary civilization, what physicists call a type 1 civilization. This civilization becomes the greatest transition in history, making a significant turn from all civilizations of the past, which have impacted culture, religion, business, trade and commerce, entertainment and other activities, and even war. Scientists believe that we will attain type 1 civilization within one hundred years from now (ibid.). Kaku, however, pointed out the double-edged character of science—science creates as many problems as it solves. We observe two competing trends in the world. First is the creation of a planetary civilization that is scientific, rational, tolerant, humane, and prosperous, and the second is anarchism, irrationality, inhumanity, and ignorance that rip the fabric of our society.

Unfortunately, still today, we have the same sectarian, fundamentalist, irrational passions of our ancestors who lived one hundred thousand years past. To be optimistic, we must make transitions from being passive observers of nature's beauty and beasts to being the craftsmen of nature and to becoming the master of it and, of course, to becoming the conservators of nature (ibid.). We hope that we can use the tools of science with our talent, wisdom, and rationalism by taming the irrational acts and barbarism that we inherited from our ancient ancestors. The twenty-first century is an exciting time as science and technology have opened up the world that we only dreamed in the past. The future of science, with its challenges and opportunities, has a genuine prospect. With no surprise, we are going to see more discoveries about nature in the coming decades than in all human discoveries combined as witnessed.

Science has given us a lot; it has taken us from the depths of darkness to the threshold of the stars. Science is unbiased, fair, and neutral. As said earlier, science is a double-edged sword, but how this powerful weapon can be used all depends on the wisdom of users. Einstein said, "Science can only determine what is, but not what shall be, and beyond its realm, value judgments remain indispensible." We witnessed the massive destructive side of science during WW I and WW II and the killing of hundreds of thousands of people by poison gas, automatic firearms and machine guns, and the atomic bomb. Nonetheless, science helped rebuild humanity above the ruin of wars and made peace and prosperity for billions of people. It suggests that the true power of science enables and empowers mankind to do better. Science helps us innovate, create, and endure the spirit of humanity, and at the same time, it helps minimize our deficiencies, because with science, we can organize knowledge in a systematic manner (ibid.).

There is no alternative of science for the future we comprehend. What kind of future we want to live in depends on how we use science. The future is ours to create, and it is only possible if we use science intelligently and productively. Nothing good can be produced without the proper use of science. Scientists are the ones who work diligently to innovate and create new things for the future. Kaku considers the future as freight trains that roll on the tracks, and behind the train are the works of thousands of scientists who create the future in their laboratories. From the whistle of the train we the sounds like biotechnology, artificial intelligence, nanotechnology, telecommunications, and the advanced technologies we discussed earlier. Although some old people can't learn this advanced stuff of science, the younger, energetic, and ambitious people can easily get on the train and navigate. I am of the opinion that the people, especially the younger of this century, will take the challenge using the magnificent tools of science wisely and with compassion (ibid.). Here's one important note of caution, though: science must not be used outside the paradigm of humanity, because science without humanity is violent.

So how would a day of a person look like in the next century? Say, the day is January 1, 2100. You wake up from bed when your wall screen lights up and a face appears on the screen. It's Caroline, the software program you loaded. She cheerfully says, "Bill, wake up." Slowly, you get up and head off to the bathroom. What do you see while you wash your face? Hundreds of hidden DNA and protein sensors in the mirror, toilet, and sink go into action. They analyze molecules you emit in your breath and bodily fluids and check if there is a slightest hint of any disease at the molecular level (ibid.). Well, it's simply unimaginable and stunning. It's a dream that you have never thought of. To me, it sounds interesting, and I love it. My grandchildren's children will have this kind of life, for sure.

Chapter Four Conclusion

"Transhumanism is a class of philosophies of life that seek the continuation and acceleration of the evolution of intelligent life beyond its currently human form and human limitations by means of science and technology, guided by life-promoting principles and values,"[158] said Max More. He rightly defined what transhumanism is. Let us have a close look. Transhumanism, in fact, is the extension of humanism because it involves humans who are dynamic in nature. We, as human beings, are not perfect. This is a fact. But we can improve human conditions by exercising rational thinking, freedom, tolerance, democracy, and taking care of other fellow human beings. At the same time, we can use various means to improve the human organism. This idea goes at par with the most basic theory that the human race can evolve beyond its current physical and mental limitations by means of science and technology. Transhumanism is about becoming "more than human" or "human plus" in all aspects of our lives—body, mind, and spirit.

We know that the future is impossible to predict as it is unborn. But this is not going to stop people from thinking and trying to explore what the future would look like and bring something new for mankind. Not only are we hoping that the technologies we invent, the social habits, and the ways we think will push us to move forward for a better existence, but we are already having a technologically better life. Kyle Munkittrick (2011)[159] pointed out the kinds of technologies that are really more important than the less important ones. He considered the indicators of transhumanism in response to how technology changes the definition of "normal" humans. He proposes several technological changes as indicators that have transformed humans into transhumans. They are prosthetics are preferred, better brains, artificial assistance, amazing average age, and responsible reproduction, among others.

Prosthetics are preferred. The usages of prosthetics and implants for organs and limbs are as good as or better than the original. Some of them are voluntary amputation, cochlear or optic implants, bionic limbs, and artificial organs, which are readily available.

Better brains. There are three ways we can improve our cognition—cognitive-enhancing drugs, genetic engineering, or neuro-implants / prosthetic cyberbrains.

Artificial assistance. Artificial intelligence (AI) and augmented reality (AR) are integrated into everyday personal behaviors. In the same way, Google search and *Wikipedia* changed the way we research, and, remember, AI and AR can alter the way we think and interact.

Amazing average age. The main objective of health care is that people live the longest, healthiest lives possible. Whether that happens because of nanotechnology or genetic engineering or synthetic organs is irrelevant. What matters is that, eventually, people will age more slowly, will be healthier for a larger portion of their lives, and will live beyond the age of 120 or so. Society will treat aging as a disease. Whatever it is, when the average expected life span exceeds 120, we would think that the conditions for transhuman longevity have arrived.

Responsible reproduction. Having children will be planned exclusively in the light of responsibility. Human reproduction is not procreation in the true sense of the term. Procreation implies planned creation and conscientious rearing of a newborn life. No one should be allowed to have a deformed or abnormal birth of a child, and when found to have that, the parents abandon the baby through abortion. The newborn child deserves better. Responsible reproduction will involve better birth control for men and women. Abortions will be reserved for rare accidental pregnancies or those that threaten the life of the mother. Those who choose to reproduce will do so with the help of assisted reproductive technologies (ARTs), ensuring pregnancy is quite deliberate. Genetic modification, health screening, and, eventually, synthetic wombs will enable the child with the best possibility of a good life to be born. We will procreate transhumans at the situation when global births stabilize at replacement rates, ARTs are the preferred method of conception, and responsible child-rearing is highly valued more than biological parenthood.

Some of these technologies do overlap with the others we have outlined in this chapter (pages 144–150) that can bring significant changes and transformations on human lives. While thinking of all these technologies, my mind resonates with what William Shakespeare said: "We know what we are, but we know not what we may become." This is very much true; we know not about the future. But it is quite certain that we are going to know the future of mankind when technology will bring overwhelming change and transformations on our lives. Transhumanists look forward for technological progress to happen not just five or ten years into the future but twenty years, thirty years, and beyond. The history of transhumanism shows the progress made by the scientists during the past half century—the rapidly increasing field of science and radical technology such as robotic implants, prosthetics, cyborg-like enhancements, and so on brought into our experience. Transhumanists think that the use of medical and microchip implants in the brain and in various parts of our body is expected to be a reality in the coming years.[160]

Zoltan Istvan, a transhumanist, believes that we can transform the United States from a war-prone military industrial complex into a scientific and educational industrial complex. This doesn't mean giving up a strong economy. Instead, it means turning bomb factories into medical research labs; turning prisons into schools, colleges, and universities that offer free education; and diverting spending on wars into spending on science and technology. I am going to recommend the same what Zoltan has suggested—instead of fighting trillion-dollar wars around the globe, it's wise to fight against cancer, diabetes, heart disease, aging, and even death.[161]

Many will argue that the transhumanist ideals and propositions will likely lead to a large disparity in the extents of biotechnological modifications between individuals and communities. People will likely feel threatened by those who are seen as different from themselves, leading to the potential of discrimination against both the enhanced and the unenhanced. This is because everyone will not be able to afford enhanced technology. Freeman Dyson (an American theoretical physicist and mathematician) thinks as follows:

> *The artificial improvement of human beings will come, one way or another, whether we like it or not, as soon as the progress of biological understanding makes it possible. When people are offered technical means to improve themselves and their children, no matter what they conceive improvement to mean, the offer will be accepted. The technology of improvement may be hindered or delayed by regulation, but it cannot be permanently suppressed. It will be seen by millions of citizens as liberation from past constraints and injustices. Their freedom to choose cannot be permanently denied.*

He reminded that in the United States, the freedom to pursue enhancement technologies is effectively guaranteed under the Constitution (life, liberty, and the pursuit of happiness).

However, the transhumanist vision of a transformed future for humanity has gone unchallenged. There are supporters and opponents. Transhumanism has been described by Francis Fukuyama (an American political scientist, political economist, and author) as *"the world's most dangerous idea."* On the contrary, a proponent, Ronald Bailey (an American libertarian science writer and author), counters Fukuyama's view by saying that it is the *"movement that epitomizes the most daring, courageous, imaginative, and idealistic aspirations of humanity."*

In the midst of arguments and counterarguments, there is a view that suggests a fundamental flaw with all utopian (transhumanist, in this case) thinking, which fails to understand the darkness, fears, and unpredictability that lie in the human psyche. Experiences in the twentieth century uphold the power games that teach us to *"beware the power of utopian dreams to enslave, destroy, and demean, rather than provide the promised justice, freedom, and human flourishing."*[162] This is a genuine ethical debate that continues, but I would rather agree on what Ronald Bailey has said. I wish that I will live long enough to see some of those imaginative and radical applications of transhuman technology become a reality.

It is estimated that from the start of humanity until the present day, 107 billion humans have lived "or are" living, with a billions more to come, so on behalf of those who have died, those who now live, and those humans who will live but are not yet alive, we have both a self-interest "and" an empathetic reason toward all, for ensuring that all of the voices of the dead are heard, that those alive don't die, and that those who are to be born have a future of greatness, health, life, and never-ending evolution to engage into the coming infinity with vigour, strength, love, and passion. We are the Transhumanists. We are the Singularitarians. Bringing life, light, and knowledge to the universe . . . and beyond . . . and the never ending to learn and create with joy and abundance, is our goal and our destiny as a human species, as a Transhuman species, and as a Singularitarian species. This is our destiny, and we embrace it. We exist to bring light to the darkness of the universe (Kevin George Haskell).

CHAPTER FIVE

Summary and Conclusion

Summary

The book *Thinking Outside the Box* is informative. It is captivating and futuristic in analyzing the development of religion, science, and philosophy. Summarizing the book, it is thought not to leave out the messages the book intends to give; they are so captivating. As a result, the summary became much longer than expected.

A religion is like a box. It forbids thinking and acting beyond what the religion edicts. It is one's choice to belong to a religion, though in most cases, a religion is drummed and imposed into one's brain at an early age, the author correctly says. It is, nonetheless, the right of an individual to have a critical view on religion. This is where *Thinking Outside the Box* comes in.

Introduction

The primitive men faced many challenges (enough food, injury or disease, and natural calamities like thunder, lighting, volcanoes, floods, etc.). They sought for security. They thought and created God or gods to give them the courage and strength to deal with life's adversities. The god idea was the response to fear, frustration, insecurity, and other crises and is central to a religious belief. Different societies, however, shaped their god(s) in different ways and gave them different names. They have separate scriptures that laid the rules for "dos" and "don't dos." The gods were taken to be supernatural, directing things from above.

On the other hand, humanity evolved with the expansion of human knowledge, experience, and wisdom. Humanism, as such, is the progressive philosophy of life that aspires to the greater good of mankind and all around it with all attributes of humanity. It promotes a world where violence and fear are not the means to achieve ideals and goals. It suggests liberty (i.e., freedom for all, freedom to peacefully affirm and practice a faith, freedom from religious coercion, and freedom to peacefully reject a religious faith). As Bertrand Russell said, "There is no reason to believe any of the dogmas of traditional theology and, further, that there is no reason for believing that they were true. Man is free to work out his own destiny. The responsibility is his, and so is the opportunity."

Our main purpose is to unite people rather than divide them. More and more people are thinking of humanity as an alternative to religion because it rejects dependence on faith and the supernatural—the sky god who controls our lives from the sky. On the contrary, humanists take control of themselves with the responsibility to lead ethical, moral, and dignified lives of individual and all others' fulfillment, wishing, aspiring, and working for the greater good of mankind and all other lives on earth. People help people—this is what humanism teaches us. And, of course, in reality, human guidance is the guidance we follow through in our daily lives.

In the religious world, there is a belief in heavenly afterlives, but many people are skeptical about it. From this came the idea of transhumanism—philosophies of life that seek the continuation and acceleration of the evolution of intelligent life beyond its currently human form and human limitations by means of science and technology, guided by life-promoting principles and values, said Max More. Transhumanism is a dynamic interplay between humanity and the acceleration of technology that is expected to lead to the transcendence of human species by creating new possibilities of and for human nature (Julian Huxley). Transhumanism, therefore, transforms static ideas of humanism as human beings would evolve by themselves. Even in Darwin's perspective, it is conceivable that humans are not the end of evolution but the beginning of a conscious and technological evolution in modern time.

Transhuman technology allows us to upgrade ourselves in terms of our physical strength and power, our intelligence, our cognitive and emotional senses, our longevity, and any other physical and mental capacity. It is not a distant matter that is far away from happening. Rather, there are plenty of technologies that are available right now that are transhuman (such as magnetic implants, implanting chips and electrodes, night vision drops, virtual reality, augmented reality, etc.). Transhumanism is not just about technologies. Our lives are not complete by adapting technology only. We need social and political establishment and progress, along with technological and biological advances, for transhumanism to become a reality.

Religion

The book traces the origin of religion (historical roots, biological origin, neurological origin of religious beliefs, the meme theory, the psychological theories of religion, intellectual theories, and social functional view of religion). All these theories lead to the worldview that people want to understand themselves and the world in which they live. This quest for knowledge is ultimately a spiritual endeavor, as it entails searching for meaning beyond the self, thus transcending the self. Traditional religion is primarily a search for security and not necessarily a search for truth. Religion is what we so often use to bank the fires of our anxiety. That is why religion tends toward becoming excessive, neurotic, controlling, and even evil.

For many centuries, it was believed that God demanded sacrifices, that he was pleased when parents shed the blood of their babes. Afterward, it was supposed that he was satisfied with the blood of oxen, lambs, and doves and that in exchange for or on account of these sacrifices, this god gave rain, sunshine, and harvests. It was also believed that if the sacrifices were not made, this god sent pestilence, famine, flood, and earthquake. During all these adversities and disasters, it was believed by all religious peoples that this god heard and answered prayers, forgave sins, and saved the souls of true believers. Now the questions are, Was religion founded on any known fact? Does such a being as God exist? Was any prayer ever answered? Did any sacrifice of babe or ox secure the favor of this unseen god? Did an infinite god create the children of men? Why did he create the intellectually inferior? Why did he create the deformed and helpless? Why did he create the criminal, the idiotic, and the insane?

If this god exists, how do we know that he is good? How can we prove that he is merciful, that he cares for the children of men? If this god exists, he has seen millions of his poor children plowing the fields, sowing and planting the grain, and when he saw them, he knew that they depended on the expected crop for life, and yet this good god withheld the rain. He saw the people look with sad eyes upon the barren earth, and he sent no rain. Do we prove that this god is good because he sends the cyclone that wrecks villages and covers the fields with the mangled bodies of fathers, mothers, and babies? Do we prove his goodness by showing that he has opened the earth and swallowed thousands of his helpless children or that, with the volcanoes, he has overwhelmed them with rivers of fire? Can we infer the goodness of God from these facts? God did not make all men alike. He made races differing in stature, color, or even intelligence. Was there goodness? Was there wisdom in this? Why should we say that God is good?

The power works for righteousness, but what is this power? It is the accumulated experiences of the natural world by which we get power and force for righteousness. For example, a child charmed by the beauty of the flame grasps it with the hand. The hand is burned, and after that, the child keeps its hand out of the fire. The power that works for righteousness has taught the child a lesson. This power and force are not conscious, not intelligent. It has no will, no purpose. It is a result. Religious people have tried to establish the existence of God in relation to morality and conscience. They insisted that the moral sense—the sense of duty and obligation—was imported, not produced by men, but from God it comes. Man is a social being. We live together in families, tribes, and nations. The members of a family, of a tribe, of a nation who make happiness of the family, of the tribe, or of the nation are considered good members. They are praised, admired, and respected. They are regarded as good— that is to say, as moral. The members who add to the misery of the family, the tribe, or the nation are considered bad members. They are blamed, despised, and punished. They are regarded as immoral. The family, the tribe, the nation create a standard of conduct of morality. Conscience is born of love, and the sense of obligation, of duty, was naturally produced. There is nothing supernatural in this. Among savages, the immediate consequences of actions were the only motive. As people advanced, the remote consequences are perceived. The standard of conduct becomes higher. The imagination is cultivated. A man puts himself in the place of another. The sense of duty becomes stronger, more imperative. Man judges himself. He loves, and love is the commencement, the foundation of the highest virtues. He injures one that he loves, and later comes the regret, repentance, and sorrow, and conscience works. As we see in all this, there is nothing supernatural.

Religion is man-made and has nothing to do with spirituality. But spirituality is native to the human being, and we are constantly in contact with that which we cannot measure in physical terms. The belief in a higher power or God does not require religion. Religion is a man-made institution developed for the control of mass people by a smaller group of people. The human race has learned how to communicate the rules and principles of morality and ethics without religious interference. We must learn how to use them vigorously for the greater good; else, our current civilization will collapse or fade away.

From a wider sense, nature has no intelligence, has no purpose. It sustains without intention and destroys without thought. Humans have some intelligence, and intelligence is the only lever capable of raising mankind. They should use it. They should be more intelligent, more conscientious, and more driven by reason than being more passion driven by impulse. Criminals, tramps, beggars, and failures are on the rise. The prisons, jails, poorhouses, and asylums are crowded. In all these, religion is helpless. The tide of vice is rising. The war that is now being waged against the forces of evil is hopeless as the battle of the fireflies against the darkness of night. But there is one hope. Ignorance, poverty, and vice must be stopped in the larger population. This cannot be done by moral persuade. This cannot be done by talk or example. This cannot be done by religion or by law, by priest or by hangman. This cannot be done by force, physical or moral. But there is one way by which it can be done—science.

Religion is slavery; it can never reform mankind. Come out of religion's box to leave the forts and barricades of fear, to stand erect and face realities of life's challenges with courage. It is far better to forget all gods, their promises, and their threats. Instead, we elevate our inner sense to do useful things, reach with thought and deed the ideas of our brain, find the subtle threads that join the distant past with the present, take burden from the weak, defend the right, increase knowledge, and nurture and develop the brain. This is real religion. This is real worship. This is to walk outside the box. A religious person is devout of superpersonal objects and goals that neither require nor are capable of developing rational foundation.

Science can only be created by those who are thoroughly imbued with the aspiration toward truth and understanding. This kind of feeling springs from the sphere of religion. Generally, people think in the possibility that the rules, regulations, and systems required for the world of existence are rational. Scientists have this profound understanding. This is what made

Einstein express the view that science without religion is lame; religion without science is blind. Einstein also added by saying, "I do not believe in a personal god, and I have never denied this but have expressed it clearly. If something is in me that can be called religious, then it is the unbounded admiration for the structure of the world so far as our science can reveal it." Einstein also said, "God does not play dice," which is cited by religious people, and said that he believes in God. It is, however, to be cautioned that how one interprets it. He perhaps referred this to the probability theory of randomness.

Charles Darwin's explanation is that Einstein used God here in a metaphoric sense. More specifically, the statement "God does not play dice" refers to the randomness of physical phenomena at very small scales (i.e., field of quantum mechanics). Einstein didn't like the idea that there is no principle for the way tiny particles like electrons, neutrinos, and photons behave, and that was the problem; he eventually didn't manage to unify his theory, as his theory of special relativity didn't work when it came to these tiny particles. So he used to say, "God does not play dice." But now, according to Stephen Hawking, we know that God does play dice and that sometimes he throws it someplace dark so that we can't see it. This has to do with the randomness of some physical principles in quantum mechanics. In general terms, Einstein didn't like randomness of things at all. So this quote represents Einstein's opinion on quantum mechanics, which relies heavily on probabilistic models of phenomena, which is in gross defiance of classical physics. Einstein was a strict determinist, which means he thought nothing in the universe happens by chance. This is summed up by the expression "God does not play dice" (Dawkins 2006).

Before Einstein, Spinoza defined God as God equals nature and man as the mode of God (one of infinite ways in which God is expressed). The idea that God watches out for us is an error. Spinoza's cool, indifferent god differs from the concept of an anthropomorphic, a fatherly god who cares about humanity. Interestingly, there are striking similarities between Vedanta and the system of Spinoza that "the Brahman, as conceived in the Upanishads and defined by Sankara, is clearly the same as Spinoza's 'Substantia.'"

Einstein was asked in a telegram by Rabbi Herbert S. Goldstein whether he believed in God. Einstein responded by telegram, "I believe in Spinoza's god, who reveals himself in the orderly harmony of what exists, not in a god who concerns himself with the fates and actions of human beings." Richard Dawkins commented that, realistically, religion is founded on local traditions and by private revelation rather than any pragmatic evidence. In the light of history of religions, there is a progression from primitive tribal animisms through polytheisms to monotheisms, such as Judaism, Christianity, and Islam.

The book summarizes the major religions (Hinduism, Judaism, Buddhism, Christianity, Islam, Sikhism, Taoism, Confucianism, and Shinto) and discusses the religious wars (nonviolence wars, just wars and holy wars, killings in religion, honor killing), child marriage, and nonreligious affiliations and shows that the last nonreligious affiliations are on the rise, especially in societies where better social justice and moral development have taken place. In fact, nonreligious affiliations (more than one billion people now in the world) occupy the third-largest place after Christianity and Islam. It, however, concludes that religion will never go away, at least in the foreseeable future. Religion, whether it's built in and maintained through fear or love, reward or punishment, has been successful at perpetuating itself. We need comfort in the face of pain and suffering, and anguish that gnaws the heart. We think that there's some immortal, omnipresent, omnipotent, omniscient, and invisible being there above us to look after us when needed. When we face an ecological crisis, a global nuclear war, or an impending comet collision, the god would definitely emerge as the savior as believers understand.

The challenge of our time is to intelligently coordinate and unify the realms of science, philosophy, and religion so that we may move toward a greater comprehension of total cosmic reality, which can benefit individual lives as well as our entire civilization. Therefore, we need a balanced understanding of the true meaning and value of our individual, interpersonal, and social lives, as well as our place on earth and in the cosmos—a viewpoint that integrates science, philosophy, and religion. We need science to enrich our knowledge and understanding about the laws of energy, matter, and our physical environment so that we may apply that knowledge to live better in our material world. We need philosophy to teach and help us how to think more clearly and to reasonably discern the universe of things, meanings, and values. Bertrand Russell mentioned (in his book *Wisdom of the West*) that the branches of knowledge have borders beyond which there is unknown and that when one passes from known (i.e., science) into unknown, there is speculation. This speculative activity is exploration, which is, among other things, what philosophy is. He further said that there are two ways to know the unknown—one is to accept what people say that they know on the basis of books, mysteries, or other sources, and the other way is to go out and look for you. This is science and philosophy. We need religion to guide us to lead our lives, to know our relationship with God and one another, and spirit values so that we may become increasingly God-conscious, live in harmony with one another, and make progress toward our mutual divine destiny. However, one can argue that the philosophy of religion is neither ceremony nor ritual, nor going to the church, temple, mosque, or synagogue, but an inner experience that finds God everywhere in nature. Here is the question, which kind of religion do we need for mankind?

There has been a time, especially during the Renaissance, when leading thoughts were directed to an integration of science, religion, and philosophy (for example, Francis Bacon, Galileo, Copernicus, and Newton). In the eighteenth century, around the time of Enlightenment, a split between science and religion started to develop. Some thinkers advocated that reason and science would be the means by which people could gain better knowledge and understanding of the universe. A movement started to eliminate irrationality, superstition, suppression, and tyranny in the society. Eventually, religion began to be undermined, and a process of secularization started to be established. Much later, the celebrated cosmologist Stephen Hawking, in his books *History of Time* and *Grand Design*, described how to eliminate God as the primary cause of existence. An evolutionary biologist and a strong proponent of Darwinism, Richard Dawkins promoted a materialistic and atheistic viewpoint. The rise and the power of scientific rationalism (like secular humanism) were seen by religious thinkers as threats to religious beliefs.

Deep into religion, there are two distinct divisions: (1) the outer exoteric traditions (mysteries, myths, fables, rituals) that cannot reform mankind and (2) the inner esoteric mysteries that deal with matter that is harder to understand (i.e., things are not as they appear, experiences that lie hidden, and this is what we experience as God). This resonates with something related to the subject of quantum physics, which states that the entire universe and everything in it are entangled and consist of a single unified "oneness"—the god. Quantum physics tells us about the behavior of particles at the subatomic level, explaining the unity and inseparability of all existence, which, in turn, allows us to justify the validity of the claims that we are part of God and the universe. Looking at it through this light, we seem to experience the oneness that quantum physics is suggesting is the true nature of things. That means all encompassing oneness that may be called god. This idea has resemblances to Spinoza's god. He thought that everything that exists is God, but he did not hold the converse view; that is, God is no more than the sum of what exists.

Perhaps there would more truth to be revealed with the progress of science resulting well-founded convergence between science and religion. But we must keep in mind that this is not about those aspects of religion (that is, the outer mysteries or exoteric traditions such as the rituals, regulations, fairy tales, and fantasies) that can be substantiated by science. It is rather the inner mysteries, the esoteric truths, and the mystical heart that exist in all of the world's religions that can be thought of. Only within the inner mysteries can one discover a set of beliefs that is common to the esoteric traditions of all of the world's religions. It is these inner mysteries that scientific discoveries are confirming, and it is these inner truths of religion that may help science reconcile. This, in turn, may help people understand and integrate some of these discoveries with humans' quest for knowledge.

Materialism supposes that the nature of existence is based on matter, whereas the main tenet of idealism is the belief that all existence is consciousness. Consciousness is related to the descriptions of God's attributes because God is universally described as immanent and within us, which is in all of the world's great faiths. Thus, we can say that consciousness is also immanent and within us. God is also universally described as transcendent—that is, independent of and beyond the realm of matter. In this sense, consciousness is also transcendent and somehow above and separate from the material world. This is the area of inquiry where materialist philosophers and neuroscientists have been working on but were not successful in explaining consciousness in the real term. The whole idea perhaps is the other way around; instead of consciousness reducing to a physical brain, it is more probable that the entire physical universe reduces to consciousness.

Humanity now begs to answer a fundamental question of how we can govern ourselves. People have formed beliefs about the set of unknown elements of the universe. Such beliefs are called probability beliefs in mathematics, science, and engineering. In the probability theory, we measure the unknown based on some perceptions and hypotheses, but we can call this belief in religious term about the uncertain elements of the universe, which is a rationally defined doctrine, or we may call religion. So the probability theory is a kind of belief, and the formation of belief process about the unknown elements is common to all fields including religion. However, this common process does not automatically produce a unifying philosophy because of the incoherence in beliefs about God across religions. The atheists, scientists, and those who do not associate with any religion do not accept God. The challenge here is to articulate a unifying philosophy that rests on presenting God in a way that will be acceptable to the people of all religions as well as to the scientists, atheists, and nonreligious people.

Scientists and philosophers have tried to define a coherent god. To present God coherently, let us divide the set of all elements in the universe into two disjoint subsets: (1) elements that are known to humans called knowledge and (2) the rest of the elements called unknowable. Both are not static; rather, they are evolving dynamically over time. Knowledge is expanding. The unknowable is shrinking, but it still remains infinite. It is logical and reasonable to say that the unifying philosophy presents the unknowable set as universal God and beliefs about the elements in this set as universal beliefs or universal God because it forms the common basis of characterizing God in all the existing religions. This is unlike the existing proclamation by many that God is unknown; God cannot be unknown, says Professor Sankarshan.

A coherently rendered god should comprise the whole unknowable set of elements. Let us suppose that God is unknown (as believed by the followers of religions). Having said so, we see that an unknown god will simply be a subset of the unknowable set. But every religion also admits that God is almighty. This means God cannot be a mere subset of the unknowable. It is because such a subset will exclude some elements of the unknowable that are not within the reach of God. This is not likely to be the case. Therefore, God, as the unknown Almighty, must be the entire unknowable set. The unknowable is, thus, a coherent rendition of a universal god that can be accepted by followers of religions as well as the scientists, atheists, and nonbelievers of religions. It would, however, be logically correct to say that the longing to reach universal God in universal religion is called scientific research, perseverance, and tenacity.

Every field of science rests on a system of beliefs or probabilities about uncertain states of nature. These beliefs are used to form expectations about the uncertain states for decision-making and controlling of events in science. Contextually, we can say that the system of modern scientific beliefs is, thus, subsumed within the universal religion. We can put forward the idea of probability and belief in econometrics, a combined subject composed of economics and statistics. Any economic variable like household income can be decomposed into two parts: one that is conditional on all the information of the economist and the other comprising the rest, which is characterized as an unexplainable, random variable or the error in the model. Economists assume (form beliefs) that the unexplainable, random error follows certain theoretical properties. Some economists can have a better model where the random error (unexplainable or unknowable part) shrinks the set of uncertainty in explaining household income. Humans tend to call the random element in their household income as luck (bad or good) because they cannot explain the random draw in this error based on known factors. This is a subject related to epistemology, finding epistemic (knowledge-based) truth. It is important to know how it works and how someone foresees the truth about some epistemic reality that others cannot.

It has been established that the ability to conform is inherent in human genes. The human gene is able to store observed facts as knowledge and then verify conformity of a new discovery or claim to the stored facts to determine the truth about the discovery or claim. The gene can mutate at birth. The repertoire of memory (e.g., epistemic logic) too can mutate (not completely erased) at birth. We see that a newborn child responds through cries when the care is required. This is a genetic response to conform facts as necessary conditions for survival. As the child grows, the gene accumulates new knowledge. This is how human genes have survived. It is the gene's nature to retain epistemic element that is needed to conform to the conditions necessary for cosurvival of the humans. Perhaps this is the reason humans, since the beginning, have found commonly acceptable tenets for coexistence or cosurvival, which may be called religion.

A similar analogy may be applied in modern science to the fact that science (despite triumphing over prevailing religions) became a new religion in the context of the mathematical probability (as belief) about unknown elements of nature (uncertainties). It is the strong conviction that the epistemic facts, retained within the genes, may be dormant in some humans. Thus, these humans do not automatically know the commonly acceptable rules for coexistence of humans. As a result, these humans are likely to be subjected to harsher and uncommon rules, rituals, and social conditions to lead life. It is possible for some people to activate the dormant genes that have active epistemic elements or facts. Having activated the dormant genes, they can conform to the truth about some commonly acceptable new rule for coexistence. Since the nature of the gene is to survive, any passive element within it is most likely to be activated when the gene's carrier (human) is faced with survival challenges. As a result, the tendency to fight for survival can activate the genetic repertoire of survival instincts. The human gene has survived and is likely to survive. This makes us strongly believe and positively think that humans will be able to discover the epistemic truth about the prevailing system of moral hazard threatening the survival of humanity in the foreseeable future.

There are two controversial perceptions about the inception of human life: one is that the human gene might have evolved from the microorganisms (that is, evolution); the other is that it might have coexisted independently and survived the onslaught of violent animals on earth (that is, creationism). Whatever is consistent with the truth, the human gene will get to the fact. The reason is that the human genes have the ability to devise safeguard mechanisms like dwellings, food, defense, and a system to prevent many hazards for its carriers to coexist. Even in the religious tenets, the rules of conduct are based on a common longing to coexist. Humans have devised different religions and have transcended national, religious, cultural, and racial barriers to devise common systems of coexistence like communism, capitalism, socialism, democracy, and dictatorship, although no one is perfect in the real sense. The search for a better system for coexistence continues. Such search for coexistence will thus converge to a universally acceptable philosophy and the way of life that is based on commonly acceptable system for coexistence.

Existing world religions do not unify global population and do not bring prosperity and stability that every human inherently cherishes. Preaching that some existing religion is superior to others is not conducive to coexistence. Only a unifying philosophy of universal religion (not the existing religious beliefs but a new one) would induce humans to persevere and attain prosperity and stability. Such a philosophy of world religion vis-à-vis humanity can bring harmony among global population. Seeking new knowledge is like discovering the elements of the unknowable through a process that is known as scientific research in modern thinking. The traditional prayer to God is a process known as religious rules and rituals. Considering that both scientific research and prayer are intended to uncover the truth, we can bring them to the domain of knowledge, which is not complete and that some elements are in the unknowable set or are parts of universal God. The process of searching for the truth is the aim of science. If the same idea is true for religion, only the esoteric element of religion is for the search for the truth. Therefore, three things come into consideration: (1) the unknowable is God, (2) science is knowledge, and (3) belief about the unknowable is religion. Such transparent and coherent definitions can remove the confusions about religions and science.

Scientists establish and deduce probability beliefs about the elements in the unknowable set in their research to make discoveries. It is essential that both atheists and theists treat the probability beliefs as the religious beliefs of the scientists. Similarly, scientists should have no hesitation to treat ancient religious beliefs (esoteric traditions in particular) as probability beliefs. To form beliefs about the unknown is common to all religions as well as all branches of science and mathematics. In this context, we have a unifying philosophy (i.e., forming common beliefs about the unknowable) that can be acceptable to the theists, atheists, agnostics, and scientists. This unifying philosophy would, therefore, be universal. In every religion, God has been perceived as the unknown almighty. Thinking of universal God as the complete unknowable set is likely to avoid such confusion. The definition of universal God as the set of unknowable elements of the universe is rational and should be acceptable to people of different orthodox faiths. This definition allows the formation of rational beliefs about God (i.e., about elements of the unknowable set) to test hypotheses as knowledge expands over time. The definition of universal God is necessary to complete a coherent philosophical tenet of science, and religion is vital to remove mutual antagonism among humans and societies, states Professor Sankaran.

Many of us acknowledge the fact that an optimal determination of tenets of any religion should be based on enhancement of stability and prosperity of humankind. As we think that democracy is the best form of governance accepted by humans as optimal (though it may not be perfect), democracy should be the fundamental tenet of universal governance system (say universal religion). Similarly, amending constitutional rules of law through optimal discourse and vote within democracy is the other tenet. Professor Sankaran pointed out that Gita in Hinduism, Bible in Christianity, Quran in Islam, and such scripts in other religions were meant to be the "guiding" rules of governance. But these scripts have not been amended to incorporate the latest human wisdom. They remain unchanged for centuries and, thus, remain dogmatic, which needs to be changed. The countries that have unshackled their governance from such dogma have enhanced their prosperity and stability than those that remain shackled. Accepting the universal religion does not essentially mean abandonment of current beliefs of an individual. Rather, this is a conducive and healthy atmosphere to live in harmony among people. The universal religion gives a complete freedom to choose and to amend or refine the script through rational arguments to synchronize with human knowledge and wisdom, Professor Sankaran concluded.

As for religion, a doctrine that remains in the dark loses its effect on mankind, often with incalculable harm to human progress. Religious teachers must have the moral and the mental strength to give up the doctrine of a personal god. This means giving up the source of fear. To be noted here is that the priests, imams, and religious preachers hold the power to infiltrate in the minds of people. They should encourage the people to use those forces that are capable of cultivating the good and the beautiful in humanity itself. Religious teachers can accomplish this, and when they do, they will surely recognize with joy that true religion has been given greater dignity and made more profound by scientific knowledge as a result, as suggested by Einstein.

Scientific reasoning can aid religion to liberate mankind from the bondage of egocentric cravings, desires, and fears. Einstein emphasized that science can achieve a far-reaching emancipation from the shackles of personal hopes and desires and can help attain a humble attitude of mind toward the grandeur of reason. This appears to be religious, in the highest sense of the word. The further the spiritual evolution of mankind advances, the more certain that the path to genuine religiosity falls not through the fear of life, the fear of death, and blind faith but through striving after rational knowledge. In this sense, Einstein believes that the priest must become a teacher if he wishes to do justice to his lofty educational mission. This idea of universal religion, in fact, is closely aligned with humanism. It is the time to bring harmony among mankind (people of all religions) by following the footsteps of humanism—a way of life that is being practiced by almost 14 percent of the world's seven billion people.

David Eller wrote, *Not all people, even all religionists, are hostile to science. There are few if any religion-defenders who dismiss science as a whole. I have never seen a serious attack on atomic theory or quantum theory or gravitational theory or, on the whole sciences like meteorology or botany or germology. Religion is generally unconcerned with these sciences, because these sciences are unconcerned with the questions that concern religion. So, one of the main failures in the battle between science and religion is the tight focus on a very few bits of science and the generalization that they (mostly evolutionary theory and the sexual or reproductive sciences) are science. So, it is not a meaningful question to ask whether religion is compatible with science; it depends on which religion, which science, and what one means by compatible.*

It is an interesting point why the religionists accept the results of science in many areas but object to some areas that contradict what superstitious ancient orthodox preachers wrote in Holy Scriptures. To explain the unresolved areas between science and religion, John Loftus reproduced the arguments of anthropologist Ian Barbour, an eminent scholar on the subject who presented four ways of relating science and religion. They are *conflict, dialogue, independence*, and *integration*.

The fundamental conflict between science and religion arises from two themes—scientism (scientific materialism) and literalism (of Holy Scriptures). Nonbelievers think that in scientism, scientific theories and methods are the principal means of knowledge and that "matter" is the fundamental reality of the universe. The scriptures' literalism takes literal interpretation of the Holy Scriptures and sets the limit for science, meaning science cannot cross the limit of what holy books interpret. This is where science and religion come into direct conflict.

The people who are scientifically conscious reject the untrustworthy approach of Holy Scriptures' literalism. Neil deGrasse Tyson argued that it is not seen that a successful prediction about the physical world has been inferred from the content of any religious scriptures. On the contrary, whenever anyone used any holy scriptures or any religious documents to make predictions about the physical world, it seemed to be wrong. Here, prediction means an accurate statement about the untested behavior of objects or phenomena in the natural world.

Dialogue takes position when science and religion go side by side on some aspects and contradict some others. It is argued that religious beliefs interpret and correlate experience, whereas scientific theories interpret and correlate experimental evidence. In other words, science explains and informs, but religion reveals and reforms. This viewpoint has been explained by Donald MacKay, who said, "Both science and theology give different kinds of explanations (with different methods and aims) about the same objects. Both explanations of the same event can be true and complete on their own levels. But the methods differ greatly. Compare how an artist, poet, theologian, or astronomer might view a sunset. They can all be correct from their perspective, even if they disagree with one another. There is no incompatibility in claiming that the formation of the universe, as we know, is the result of natural processes and that the cosmos is God's creation." Each explanation is from a particular conceptual framework and can be true from the perspective of that framework.

Others have similar arguments. For example, Howard Van Till has argued that when scientists make statements concerning the origin, governance, or purpose of the cosmos, they are necessarily going outside the boundary of science and drawing from their philosophical (religious) perspectives. Theologians do the same thing. When they make statements about the geologic processes or thermodynamic phenomena or cosmic chronology, they are necessarily going outside the boundary of scriptural interpretation and getting into the domain of natural science. But this kind of dialogue between science and religion is unlikely to happen because both parties have very different views and they justify points of view from their own perspectives. Richard Dawkins thinks that it is not necessary to have dialogues between science and religion and says why we should not comment on God as scientists. A universe with a creative superintendent would be a very different kind of universe from one without. Why is that not a scientific matter? Science concerns itself with *how* questions, but theology is equipped to answer *why* questions. There are some genuine questions that are beyond the reach of science. But if science cannot answer some ultimate question, then what makes anybody think that religion can? Take a look on religious perspective. Theologians have no evidence of any natural phenomena; their tenet is only belief, nothing more than that.

Science and religion stand on two different platforms, and they are independent of each other. They have contrasting arguments based on two different methods. Science is based on evidence and deals with nature, whereas religion is based upon belief and deals with God. So science and religion are two mutually independent realms; science occupies the empirical realm of fact and theory, whereas religion deals with ultimate meaning, purpose, and moral values.

The United States' National Academy of Sciences supports the view that science and religion are independent.[163] Science and religion are based on different aspects of human experience. In science, explanations must be based on evidence drawn from examining the natural world. Observations and experiments based on scientific methods that conflict with an explanation eventually must lead to modification or even abandonment of that explanation. Religious faith, in contrast, does not depend on empirical evidence and is not necessarily modified in the face of conflicting evidence

195

but typically involves supernatural forces or entities. Because they are not a part of nature, supernatural entities cannot be investigated by science. In this sense, science and religion are separate and address aspects of human understanding in different ways.

Ian Barbour, however, suggests that science and theology can be integrated with each other. The integration model suggests that science and theology may be combined to create a more coherent view of reality. He thought that the content of science and theology can be integrated in at least three ways. They are *natural theology*, *theology of nature*, and *systematic synthesis*.

It may be argued that some of God's characteristics can be known only from revelation in scripture, but the existence of God can be known by reason alone. While Darwin argued that adaptation can be explained by random variation and natural selection, he later revised the argument as saying that God designed not the particular details of individual species but the laws of evolutionary processes through which the species were formed, leaving the details to chance. So natural theology states the fact that understanding of nature can make us think and support theology, as suggested by some theologians (Loftus 2012).

Theology of nature does not start from science; rather, it starts from a religious tradition based on religious experience and historical revelation. However, theology of nature states that some traditional doctrines need to be reformulated in the light of current science. Here, science and religion are considered to be relatively independent sources of ideas but with some areas of overlap in their claims. So our understanding of the characteristics of nature is likely to affect our models of God's relation to nature.

In contemporary views, nature is understood to be a dynamic evolutionary process characterized by both law and chance. This suggests that the natural order is ecological, interdependent, and multileveled. These characteristics can modify our thinking of the relationship between God and humanity to nonhuman nature, which, in turn, affects our attitudes toward nature. Biochemist and theologian Arthur Peacocke supported the harmonization of religious belief and scientific theories. He is willing to reformulate traditional beliefs in response to current science by discussing how chance and law work together in cosmology, quantum physics, nonequilibrium thermodynamics, and biological evolution.

Ian Barbour (2003) suggested that through systematic synthesis, a systematic integration can occur if both science and religion contribute to a coherent worldview elaborated in a comprehensive metaphysics. Panentheism, or process philosophy, asserts that the universe is the creation of God in his entities, which is defended by Paul Davis, Alfred North Whitehead, and Charles Hartshorne. According to them (as for evolutionary thinkers), nature is a dynamic web of interconnected events, characterized by novelty as well as order. Process thought states that the basic constituents of reality are not two kinds of enduring entity (i.e., mind/matter dualism) or one kind of enduring entity (materialism) but one kind of event with two aspects or phases. All integrated events have an inner and an outer reality, but these take very different forms at different levels. God elicits the self-creation of individual entities, thereby allowing for freedom and novelty as well as order and structure (Loftus, 2012).

John Loftus pointed out that it is immensely difficult to put into practice. Even if the integration takes place, religious beliefs are the ones that need to be integrated with science and not the other way around, the reason being that scientific theories can be tested empirically in a dialectical manner, whereas religious faiths cannot be established by mutually agreed, reliable test. There is knowledge gap to understand God. Christian philosopher Robert Larmer holds the idea that it is not anything wrong with arguing from the gaps in our knowledge to the existence of God. But this argument may not be credible considering the progress of science that has led us to close many gaps of our knowledge in the existence of God and that divine God will no longer be needed to lead life. We may argue that the development and continuous progress of science and technology down the centuries have changed or would likely change theistic mind. John Shook echoes the argument that theologians are wrong so far and said the following:

Every generation of theologians who made careers predicting that science could never explain something had impressive stances right up until the time when science did explain it. Examples over the last ten centuries are assertions that science would never explain nonlinear motion, the nature of the stars, magnetism, light, how organisms use air, the age of the earth, the diversity of life, and the transmission of life in reproduction. Upon these mysteries and many more, believers constructed elaborate philosophical and theological systems to do what they thought science could never do. But in time, physics, astronomy, chemistry, geology, biology, and genetics have satisfactorily explained all these things. The history of such metaphysical speculations trying to outmatch science is lettered with exploded and abandoned systems of thought. It has always been a bad bet to bet against science.

The origin of religious beliefs was prescientific. The people of those times had little or no understanding of the natural processes such as sunrise, rainfall, earthquake, thunderstorm, and so on. Those prescientific people believed in a magical world. They performed rituals and prayed to God for the things they desired. Naturally, whatever they could not explain was attributed to God. On the contrary, it is science that discovered the natural phenomenon and solved the mysteries. It is science that has opened our eyes and rewarded the knowledge of nature and life and theories regarding many of natural phenomena. What is the alternative of science and of humanity? Blind faith is just for personal satisfaction without thinking in the matters and events whatsoever but is not the rational way of pursuing our lives. The difference between blind faith and reason is just like—it is easy to believe than to think.

How do we bring scientific knowledge into religion? Let us examine. Ancient religions used myths to explain the mysteries of nature by making fables and stories. In modern religions, the ancient ideas took new forms in the holy books and scriptures, which are believed to be the words of God. But those scriptures were written by people who had only primitive understanding of the universe. They could not grasp that the earth circled the sun. They had no understanding of the solar system that was a part of a much larger galaxy among billions of galaxies. With the knowledge of modern science and technology, we now know that what the writers of holy books wrote about was beyond their understanding.

Two elements of science may be looked into aimed at opening discussion on the subject in question. The first is that nothing is fixed in the world and in the universe; rather, all things are in motion, and all things are evolving, including human beings. Human beings are not the product of a onetime creative act by an all-powerful god; human beings are the result of a continuous long process. Charles Darwin's work on biological evolution pushed forward critical discussion in Christian theology. Others moved the discussion to theology, which suggests that everything is changing in a dynamic process of creative advance that will never end. This includes God as well (Loftus 2012).

The second element is that there is no beginning and there will be no end. We know that scientists are now analyzing the outer limits of space in terms of billions of light-years, which are still expanding at accelerating speeds. Thus, beginnings and endings of universe are no longer relevant concepts, some suggest. However, in the Bible and in the Quran, there are verses about beginnings and endings. The concept of the creation of heaven and earth in seven days relates to end-time theology. But in the light of modern science, this kind of thinking does not make any sense, so it is irrelevant.

There is no clear answer to the question on what kind of religious explanation can relate to science that embraces life that is never static and is always in a process of change. The matter of fact is that science continues to progress, but there are still lots of unknowns that scientists have to search answers for. On the other hand, religion needs to create a widely acceptable environment in which people can find the complete and meaningful life. In religion, the end justifies the means. Science is an evergreen area of knowledge where the unknown becomes known in a gradual process. One may argue that science is in its infancy; it is never ending but not complete. There is no denial of this fact. Religions need to be open, dynamic, creative, and enjoyable, rather than static, compulsive, and fearful to the lives of people. Only then is there a possibility that scientific knowledge can flourish in religions. Rev. Howard Bess (2015) hopes that life will be fun, enjoyable, and rich when religious people and scientists are on the same dance floor. Does anyone hope for this to happen? If so, when? These are the open questions.

Reverend A. Powell Davies has explained this subject in a different perspective. According to him, science must enter into the field of religion, or religion must get into the field of science. He states that religion must be liberal, may be termed as liberal religion, and it must maintain the open-mindedness toward future discovery. It should not be restricted by a creed. In other words, when new knowledge comes, religion must accept it and take the consequences of it. This suggests that religion should accept the advancing truth, accommodate all knowledge and wisdom, and always try to know what experience justifies. Only with this kind of liberalism in religion can science meet with religion. Only this kind of religion can keep the door open to scientific advancement without barriers imposed by religious doctrines. This, in turn, suggests that when scientists go along with religion without abandoning their scientific disciplines, they cannot accept a traditional creed as binding. Scientists want a free field without church's interference and a free and open religion so that they can openly pursue their intellects. It would be rejoicing if traditional denominations would allow this freedom not only to scientists but also to others.

Reverend Powell Davies thinks that scientists do have a need of religion, its basic faith, its moral responsibility, its deeper insights, its wisdom, and its inspiration. Religions need to be genuine and pragmatic. Let us hope that when science and religion meet, their knowledge and resources are mixed together in an honest approach for the common good. What is essential is that religious scholars and preachers truly embrace the scientific methods and the results, and in the same way, scientists take religion to bring into its domain whatever possible through open discussion.

He brought forward what Sir James Jeans said from the viewpoint of science, "The universe begins to look more like a great thought than like a great machine." This statement comes from liberal religion's perspective but not as an endorsement of the Apostle's Creed. Sir Arthur Eddington stated, "The idea of a universal mind . . . is a fairly plausible inference from the present state of scientific theory." Let us consider that this statement is not in the least the same thing as confirming the doctrine of the Holy Trinity. Prof. Arthur Compton said, "There is something of a nonphysical nature which controls the action of the atom." Let us consider that by this statement, he has not declared his adherence to the *Westminster Confession*. Albert Einstein said that he believes in God, "the god of Spinoza." Religion should not make falsities to be true and generate fears but power in our hearts. We must break the bondage of the religious past and become liberated from closed-box thinking to the real, open possibilities of life. In today's civilized world, religion is required to accept that, like science, the fact of truth is supreme—the free truth in the open world, the truth as experience, and open knowledge prove it. Let there be the truth of the heart of people from knowledge; let there be only truth we seek truthfully that meet our heartfelt needs and desires. To this end, let science and religion meet and mingle. Again, the big question—does anyone hope for this to happen? Millions of people are of the same opinion as Reverend. A. Powell Davies (cited in *Why I Became an Atheist* by John Loftus, 2012).

One important subject is the morality—what do people think about it? Does morality depend on religion? These questions are thought-provoking and philosophical indeed. Let's see what morality is about and how we define it; we may get some insight. Common thinking is that morality depends on the existence of God. For example, some people suggest that there is no right or wrong without God or that atheists who do not believe in God can have no objective basis for their values and that their lives are entirely meaningless. Some people even think that the existence of a moral conscience supports the existence of God.

One reason why someone might think that morality depends on God is that he or she accepts, explicitly or implicitly, the divine command theory of morality, which essentially says that "morally right" means "commanded by God" and that "morally wrong" means "forbidden by God." However, there is a powerful objection to the divine command theory. This objection derives from a discussion of Socrates in one of Plato's dialogues, which are (1) conduct is right because the gods commanded it or (2) the gods command it because it is right.

With the first option, God's commands can be seen as arbitrary, in that God could have given different commands just as easily. He could have commanded us to be liars, and then lying would be right. If God says that we should be dishonest, then again, this is what we should do. Then the goodness of God can be reduced to nonsense, because if we accept the idea that good and bad are defined by reference to God's will, this notion lacks good sense. In the second option, where God commands a behavior because the behavior is right, we admit that there is some standard of right and wrong that is independent of God's will. What it means is that because it is right, God commands it is right. We might ask ourselves why (if behavior is right without the commandment of God) God should bother to command it. Is it not that the theological definitions of right and wrong are unnecessary? So adopting a theological definition based on divine command theory would mean accepting the first option but without the goodness of God, because God commands it, and that is why it is right.

There is another theory known as the theory of natural law—a system of law that is determined by nature, and so it is universal. Classically, natural law refers to the use of reason to analyze human nature, both social and personal, and deduce binding rules of moral behavior from it. Under this theory, moral judgments are based on reasons. St. Thomas Aquinas, a strong supporter of natural law, emphasized that acting reasonably is not to be contrasted with acting as a theologian. Morality is not a matter of faith but a matter of reason and conscience. Religious considerations do not provide definitive solutions to many of the controversial ethical issues that we face today. James Rachels[164] explains that under natural law, morality and religion bear the same relationship as science and religion. Science is autonomous and has its own questions and standards of truth, but religious people understand its findings in their own way, such as how scientific results in physics and astronomy may provide information about how God chooses to arrange the universe.

Religious people say that belief in a god who will punish and reward us in the afterlife on the basis of our deeds is a necessary component of moral motivation. However, there is a problem here, too, that many atheists and theists behave morally but not out of fear of punishment in the afterlife. One might argue that even if God has given all of us the ability to tell right from wrong, believers have an advantage because of revelation, where God tells the faithful how to conduct their lives. More radically, one could say that morality totally depends on revelation. But there are problems that we have to consider that count as sacred texts and what their teachings are. Which texts or oral traditions constitute God's message to us? For example, is it the Hebrew Bible? Or is it the Christian Bible? If so, which version? Or is it the Quran? Or is it any other religious sculptures? Believers justify their actions by appealing to their respective sacred books. If religions per se tell us that killing is immoral, why don't they strongly excise the relevant passages from their sacred books that ask to kill? They don't do it because not one word or passage of the holy books can be changed, because it is verbatim the word of God. Now we are back to the fundamental—we have to believe what believers decide to interpret their sacred books that clearly contain injunctions that we should be killed if we do not follow the commands interpreted by them. This is not accepted by the nonbelievers and open thinkers.

More than 80 percent of the world population adheres to some form of religious belief. It is curious to know how many believers are unaware of the history of their respective religions. Majority of them know only the stories told in their scriptures. They hardly search for the truth that lies beyond stories of scriptures. As they live in the religious box, they are devoid of the open-world concept. But humanity is more than religion, and at times of history, humanity would be better off without religion. Everyone who appreciates the good, the true, and the beautiful gifts of nature has a duty to challenge the religious superstition in every way. It is not enough to be irreligious; we must use our knowledge, intelligence, and even critique to expose religion for what it stands for. To speak rationally, it is increasingly dangerous in a society that is defined by fear. Somewhere, there are fanatics who think killing people who oppose religious views will be their ticket to heaven. We must stop those fanatics. It won't be easy, because shifting the momentum of history never is. We may not actually hasten the demise of religion (that would be too much to hope for), but we can slow down further growth and eventually slide to the bottom. As we stand today, there are only limited or zero spaces of freedom by living in a religious box. This is what we fight for—coming out of the box. Our work exposing the contradiction among religion and morality and humanity will hopefully preserve our freedom to think openly.[165]

In the earliest Western legal systems, the existence of human rights is derived from secular logic, rationality, and humanitarianism. It is found in the seventeenth-century book of Hugo Grotius—*De Jure Belli ac Pacis*. The book became famous for codifying mortality without any need of laws and divinity but based on reason and humanitarianism. Since then, human rights have become an increasingly powerful tool used in the fight against arbitrary oppression, intolerance, and unjust mob rule. We can also refer to Jack Donnelly, a political theorist who specializes in human rights and is the author of *Universal Human Rights in Theory and Practice*. He emphasized that the source of human rights is man's moral nature and that internationally recognized human rights do not depend on any particular religious or philosophical doctrine. It is to be noted that truths are proclaimed by human rights documents themselves—that *these rights derive from the inherent dignity of the human person* or in the Vienna Declaration that *all human rights derive from the dignity and worth inherent in the human person*. In the context of nonsectarian and postcultural view, the human rights come from moral understanding and thinking, but not from any particular religious or ethical philosophy. Human rights document is purely secular and a universal concept of human rights.

Prof. Victor J. Stenger pointed out the fact that religious codes and exemplars cannot literally be the origin of people's moral thoughts. These moral thoughts are remarkably similar in people with different religious concepts or without any such concepts. Even religious people's thoughts about morals are constrained by intuitions they share with other human beings more than official codes and models. Religious nobles and preachers tell us that any universal moral standards can only come from one source— their particular god. Otherwise, standards would be relative, depending on culture and differing across cultures and individuals. The data, however, suggest that the majority of human beings from all cultures and religions or with no religion agree on a common set of moral standards. While specific differences can be found, universal moral norms do exist. But what is the source of universal morals? Scholars like Socrates, Pascal Boyer, Albert Einstein, Bernard Williams, Greg Epstein, Julian Baggini, Daniel

Dennett, Christopher Hitchens, Richard Dawkins, and many others expressed their views and opinions about morality and its source, which have been discussed in chapter 1.

Humanism

Humanism is a philosophy, a worldview or life stance that focuses mainly on nature and humans and rejects the supernatural. Humanists think and accept that this is the only life we can know of and that it is our only time to live and enjoy to the fullest. It is our individual and collective responsibility to cooperate and work together in trying to find solutions to the world's problems and to preserve the planet now and in the future. The term implies not only such qualities associated with the modern word *humanity*—that is, understanding, benevolence, compassion, mercy—but also more characteristics as fortitude, judgment, prudence, eloquence, and even love. In short, humanism calls for the comprehensive reform of culture, the transfiguration of what humanists termed the passive and ignorant society of the Dark Ages into a new order that would reflect and encourage the grandest human potentialities.

The major developments in literature, philosophy, art, religion, social science, and natural science had their basis in humanism. Prof. Robert Grudin (University of Oregon, Eugene, author of *The Grace of Great Things*) and others said that the modern awareness (that is, the sense of alienation and freedom applied both to the individual and to the race) derives ultimately from humanistic sources. Apart from its skepticism and inner conflicts, the humanistic movement was heroic and remarkable in its aspirations. Humanity's moral values and programs form the basis for lives that are remembered with admiration, that we should follow and practice not only at the individual level but also in the community and state levels. The core of humanism has perhaps been best expressed by the philosopher Baruch Spinoza when he wrote, "Peace is not an absence of war. It is a virtue, a state of mind, a disposition for benevolence, confidence, and justice." This vision is a global theme that we must adapt to the realities of our times.

During the French Revolution and soon after, in Germany, the so-called Left Hegelians began to refer *humanism* to an ethical philosophy centered on humankind, as opposed to institutionalized religion. Religious humanism refers to organized groups that sprang up during the late nineteenth and early twentieth century. It centered on human needs, interests, and abilities rather than the supernatural. Humanism as a philosophy existed in Asia since ancient times. Approximately 1500 BCE, human-centered philosophy that rejected the supernatural can be found in the Lokayata system of Indian philosophy, in Iran, and in China. In the Taoist and Confucian secularism, elements of moral thought devoid of religious authority show some resemblance to the modern concept of secularism. Many medieval Muslim thinkers and philosophers pursued humanistic, rational, and scientific discussions and debates in their search for knowledge, meaning, and values. Many writings on love, poetry, history, and philosophical theology by Islamic writers show that medieval Islamic thought was open to the humanistic ideas of individualism, secularism, skepticism, and liberalism.[166]

It is also suggested that humanism, as a philosophical and literary movement, originated in Italy in the second half of the fourteenth century and spread all over Europe. As an atheistic theory, it was conceived in the seventeenth century; but as a theistic pragmatic theory, it was introduced indirectly at around 200 BC, at the time of *Vedas* and *Upanishads* in India. The latter half of the nineteenth century witnessed Hindu Renaissance, pioneered by Brahm Samaê of Raja Ram Mohan Roy and Arya Samaê of Dayanand Saraswati, finally blossoming into Vedantic Hinduism of Vivekananda. Vedantic Hinduism stresses the importance of service to the weak and the needy as its practical aspect. The salient theme is "Society is the greatest where the highest truths become practical." Humanism has undergone significant development at various levels and forms in the West, as well as in the East. But there is a basic difference between Western humanism and Eastern humanism. While the former is atheistic in content because of the conception of God as the creator, the latter is the Vedantic humanism, which is not atheistic.[167]

Renaissance and Reformation of Humanism

Humanism gives primary importance to human beings, and its outstanding historical example was the period of Renaissance humanism from the fourteenth to sixteenth centuries, rediscovered and developed by European scholars of classical Latin and Greek texts. During that time, much of the wisdom of the ancient world was lost or destroyed, in which intellectual life was dominated by religion and theology. In fact, there was little or no freedom for most people, and religious dissent, or heresy, was harshly punished. It is often called the Dark Ages. In opposition to the religious authoritarianism of medieval Catholicism, strong emphasis was given on human dignity, beauty, potential, and every aspect of culture in Europe, including philosophy, music, and arts. Humanist emphasis influenced reformation and, in turn, brought about social and political change in Europe. From the ninth century onward, important European cultural people laid the foundation for the Renaissance.

Intellectuals and thinkers were seeking ideas from intellectuals all over the world. Many Arab scholars in the areas of mathematics, astronomy, and medicine were brought to Europe. It was an era of exploration and discovery. However, the church was often hostile to new scientific ideas, which it saw as threatening. When the Polish astronomer Copernicus (1473–1543) suggested that everything in our solar system revolved around the sun and discarded the traditional idea that everything revolved around the Earth, this was opposed by the church, as many others, such as Giordano Bruno and Galileo, who publicly accepted or developed Copernicus's ideas. In England, Francis Bacon (1561–1626) developed a theory of scientific method and recommended a thorough collection of data before drawing conclusions.

The visual arts became characterized by a growing realism. The use of perspective and drawing from nature were reflected in arts, and arts gradually became more diverse rather than only religious. Famous Italian artists in that period include Uccello, who was an early user of perspective in his paintings, and Leonardo da Vinci, who was known as Renaissance man for his wide range of interest and knowledge of art. Leon Battista Alberti (1404–1472) was another Renaissance man for his wide-ranging abilities and interest regarding nonreligious ethics and a view of citizenship and the architecture of cities that was very secular and modern. His idea was that the city must provide the best possible setting for its citizens and that the architect of the city must serve the needs of man with dignity. Playwrights such as Christopher Marlowe and William Shakespeare developed a new kind of theater that was more secular, more realistic, and more interesting in human psychology and emotions. The invention of the printing press at that time was an effective means of dissemination of ideas. Aphra Behn, the first Englishwoman to earn her living by writing, wrote critically about religion and slavery.[168]

While the process and ideas began during the Renaissance in the seventeenth century with a growth in religious dissent, the eighteenth century was a period of intellectual discovery, and the dissent (religious, political, and social) became more open. A few enlightened rulers, such as Frederick the Great of Prussia, were patrons of radical writers and thinkers, fostering the growth of new ideas. Notable figures like the radical philosopher and campaigner Thomas Paine influenced the French and American revolutions that took place at the end of the century, and Mary Wollstonecraft pioneered feminist ideas in her writings. Atheism was uncommon and persecuted, but criticism of organized religion and traditional religious beliefs was widespread. Religious skepticism became more common in the eighteenth century as a consequence of the development of a more scientific view of the universe. Scottish philosopher David Hume wrote very critically about miracles and religion. In France, a group of radical and freethinking philosophers, who were highly influential, expressed their liberal, materialist, empiricist, and naturalist ideas and their skeptical attitude to religion. Their ideas influenced the course of the French Revolution. In Germany, the philosopher Immanuel Kant revolutionized the studies of metaphysics and ethics.[169]

The nineteenth century is often thought of as being a pious age. On the other hand, it was a period of skepticism and renunciation of faith for many thinkers as well. Humanist thinking developed rapidly in the nineteenth century because it was closely associated with new scientific thinking and discoveries. Darwin's ideas and new biblical research and scholarship coming from Germany provoked a crisis of faith in many Victorian intellectuals. Darwin's defender T. H. Huxley coined the word *agnostic* to describe his belief that there were things that we could not possibly know. The most notable publication of the nineteenth century was Charles Darwin's *The Origin of Species*. Published in 1859, it explained the origin of man by describing evolution by natural selection over millions of years and confirmed what many had suspected. Upon learning how life on earth evolved and realizing that there was no need for a creator, many people became agnostics. There were many nonreligious scientists who were motivated by the desire to gain more knowledge about the workings of the universe around them and to help improve the condition mankind was in.[170]

Humanism, as it was conceived in the early twentieth century, rejected the revealed knowledge, religion-based morality, and the supernatural. Fewer people in Europe were actively religious, and people were free to declare their disbelief in gods with little fear of reprisal or social disadvantage. Most of the twentieth-century philosophers worked on the assumption that morality is independent of religious faith. Because of their belief that this world is the only one we have and that human problems can only be solved by humans, humanists have often been very active social reformers. There have been huge developments in science and medicine that have affected people's lives and the way they think. As more and more people around the world got education, understanding of science has become much more widespread, and once-controversial ideas such as Darwin's theories about evolution were generally accepted. Thanks to the relatively new sciences of sociology, anthropology, and psychology, our understanding of human nature and society has developed rapidly. Many scientists were and are humanists. Albert Einstein, who worked out the theory of relativity and one of the greatest achievements of the human intellect, was essentially a humanist.[171]

Humanist Manifesto

The first *Humanist Manifesto* was issued by a conference held at the University of Chicago in 1933. It identified humanism as an ideology that espouses reason, ethics, and social and economic justice, and they called for science to replace dogma and the supernatural as the basis of morality and decision-making.[172]

The International Humanist and Ethical Union (IHEU) is the world union of 117 humanist, rationalist, irreligious, atheistic, bright, secular, ethical culture, and free thought organizations in thirty-eight countries. The following is according to the IHEU's bylaw 5.1:

> *Humanism is a democratic and ethical life stance, which affirms that human beings have the right and responsibility to give meaning and shape to their own lives. It stands for the building of a more humane society through an ethic based on human and other natural values in the spirit of reason and free inquiry through human capabilities. It is not theistic, and it does not accept supernatural views of reality.*

Humanism is about maximizing the safety, well-being, and potential prosperity of all people in our society, putting emphasis on human values, on human rights, and on humane behavior toward one another. It is a set of principles that can appeal to the whole of society regardless of their personally held religious beliefs or atheism, as it is about designing society through optimizing human collaboration around what benefits us all in this life, here and now, in the shared social space. This democratic values and principles should be the political philosophy, economic model, and vision for a better world; this is democratic humanism.[173]

Knowledge, Ethics, and Value of Humanism

To gain knowledge, humanists use reasons and experience to understand the world. And they may create the great artistic fruits of humankind to enhance their emotional palettes, deepen empathy, and enrich understanding. However, humanists reject any reliance on blindly received authority or on dogma or on any divine revelation.

A humanist can grasp ethics. They are not being the only moral subjects (other animals deserve moral consideration too). But human beings have a unique capacity for moral choice, such as acting in the interests of welfare, advancement, and fulfillment, or against it. To act right, they must take responsibility for themselves and others, not for the sake of preferential treatment in any afterlife (even if we believed in it, that motivation wouldn't make our actions good!) but because the best they can do is to live this life as brilliantly as they can. That means helping others in community, advancing society, and flourishing at the best.

And humanists are the ones who find value in themselves and one another, respecting the personhood and dignity of fellow human beings, not because they are made in the image of something else (human beings are a product of evolution, not the product of a divine plan) but because of what we are—a sentient, feeling species, with value and dignity inherent in each person. There is no reason to believe that meaning has to come from a supreme being. Humanists recognize that there is no divine plan or purpose. We make our own purposes, tell our own stories, and set our own goals. This gives life meaning.[174]

Humanism is not an ism. One doesn't "convert" to humanism like one converts to a religion. Rather, one automatically believes in humanism. It has no sacred texts, no sourcebook of unquestionable rules or doctrine, no public worship, no founding figurehead, and no structure of authority. However, humanism has certain ranges of beliefs and values that constitute a view of the world, a philosophy based on which many people live their lives. Humanism is an alternative to religion that fulfills much the same function as a religion. It is demanding but immensely rewarding. It puts a lot of responsibility on you to think for yourself, but it provides you with the freedom to do so and a basis on which to make ethical decisions. You don't have to have a religion, but if you don't have a life stance of any kind, you are rootless, without answers or purpose. Humanism provides the answers for those who can't accept religion but want an ethical approach to life, and it brings with it the inheritance of a glorious history and the promise of a better future.[175]

Humanists' beliefs are naturalistic. They believe that the universe can be explained by natural laws; many of the laws have already discovered, and many of them are yet to be discovered. Humanists have no belief in an afterlife, and death is the end and there is no survival after death. Almost all religions believe in a continued existence after this life; some of them, a reincarnation in this world; others, a translation to a different realm of existence; and sometimes, they believe in existences before this life too. Humanists think that we are not soul trapped in a mortal body: what we are resides in our bodies and brains, and bodily death means the end of the vastly intricate system of matter animated by electrochemical impulses that make us.

Humanists also believe that human beings are moral creatures. They have the capacity to think in moral terms and cannot live without morality, which is ingrained in human nature. Humanists say that biology and culture have created our moral sense. There are prosocial behaviors (such as altruism and cooperation) that are necessary for living together with others of your own species. These behaviors are the evolved mechanisms shared by all human beings. Humans have lived as social animals since millions of years before they were even fully human, and all social animals have rules and patterns of behavior that enable them to live together in a harmonious and productive manner. Humans survived and made progress with the development of language and the ability for abstract thought. They refined unwritten rules into moral values. Our instincts (which are in our brain) are the basis on which the concept of morality is built. Yes, we are not naturally, exclusively good, as some instincts are aggressive or selfish, and some are group focused, which are hostile to outsiders. But human nature has the attribute of plasticity; this is because of wrong education and experience and even of environment. Given those conditions, many people can adapt antisocial behaviors and indulge in doing wrong things. These suggest that our moral views undergo redesigned process that is built on by culture, but at the root, the morality resides in human nature, hardwired in us.

Life, as a phenomenon on earth, has no purpose. Human life has no purpose in a sense analogous to any object on earth, which has some purpose. We, as humans, have the capacity to create meaning and purpose for ourselves. What meaning we have is the fact that of our own making; that is, meaning and purpose are human constructs. We only give our own lives a purpose; that is, we adopt certain goals that seem meaningful to pursue, and we shape our lives according to our desired goals and work hard to achieve them. As a matter of fact, a meaningful and purposeful life constitutes one's individual way of defining life, which may differ from others. This is because each person is inherently different in their talents, thinking, ideas, learning, and interests. So the purpose and meaning of life differ from person to person. However, one may argue that the very purpose of life is to lead a good, fulfilled life and seek happiness.

What makes our lives meaningful and makes us happy? This seems more like personal happiness, which is, by nature, self-centered. Dalai Lama argued that happy people, in contrast to personal happiness, are found to be more sociable, flexible, and creative, and are able to tolerate life's daily frustrations more easily than unhappy people. And most important is that they are found to be more loving and forgiving than unhappy people. Robert Ingersoll, the American freethinker, also contemplated happiness in the way by saying that reason, observation, and experience have taught us that happiness is the only good, that the time to be happy is now, and that the way to be happy is to make others so. David Pollock, while explaining this issue, has mentioned that happiness is something much more substantial. Happiness is something about one's relationships with other people, and our personal happiness is inextricably tied up with that of others and is related to emotional contact. He further cited another formulation that comes from Bertrand Russell: "A good life is one inspired by love and guided by knowledge." Love and knowledge recur in humanist conceptions of the good life—love, because one's inner life and emotional life is vital and because it is the relational aspect of life, emotional fulfillment, sympathies, and affections, in relation both to others and to the natural world; and knowledge, because learning, knowing, and understanding give joy and fulfillment to life.

Humanism gives us every alternative to religion to lead our lives individually, socially, and globally. It puts a lot of responsibility on us to think of our personal lives as well for others and also gives us with the freedom to do so and provides a strong basis on which we can make ethical decisions. We don't have to be associated with a religion, follow rules, and practice rituals. However, we need to have a life stance, a guideline without which we are rootless and our lives will be chaotic and lack direction. I am fully convinced that humanism provides the guidelines and answers for those who cannot accept, in principle, any of the religions but want an ethical, moral, evidenced-based, and practical approach to life. Humanism can bring with it the inheritance of a glorious history that mankind practiced since the inception of modern man. Humanism promises a better future for mankind.[176]

How do humanists find meaning and purpose of life? Well, in all counts, the very purpose of life is to live; with this, we can add *happily*. Metaphysically, one may argue, "Life is without meaning." However, we bring the meaning to it. The meaning of life is whatever we ascribe it to be. The naturalistic view seems, to me, a realistic one: "Being alive is the meaning." We create our purposes in life and then strive to achieve them in itself, provide a sense of meaning. This meaning is derived from the good we do, such as relationship building, the quest for intellectual growth, effort for physical and mental development, the satisfaction of productive work, the enjoyment of creative or artistic pursuits, the influence we create within society and work, helping others when needed, etc. But the meaning of these achievements is not driven by any outside authority; rather, we define the meaning of life by ourselves. We can summarize this meaning as "The purpose of life is to live a life of purpose." Obviously, this statement is subjective; one has to find meaning and purpose of his or her life self-defined. "Dum vivimus, vivamus, Horace" (Since we are living, let us live well). Yet many have suggested that the principal aim of life, in most rudimentary form, is to survive to the fullest efficacy possible and to replicate one's own genetic lineage. However, the meaning of life is open to interpretation in different ways by different scholasticism, ideologies, knowledge, and education of the individual. So in a subjective sense, the meaning and purpose of life is open to interpretation.

Secular humanists tend to find the meaning of life to be the pursuit of life in abundance, happiness, pleasure, and love. Categorically, the meaning of life is composed of a number of categories such as psychological, biological, social, political, scientific, and philosophical. We can look at it ourselves through discovering who we really are and then relate the meaning of life with understanding and awareness of one's self. We must realize and pursue the best of our abilities by cultivating individual personality, strength, intelligence, knowledge, skepticism, and empathy. The meaning and purpose of life as humanists in the social purview is we need to work toward the remediation and reconciliation of social ills and conflicts, to help create a peaceful, cohesive, social inclusion, assimilation, integration, and tranquil social environment. The meaning and purpose of life is to affirm universal human rights and decency and to work toward creating a peaceful and harmonious life among populations worldwide. It is our moral responsibility and obligation to uphold the Universal Declaration of Human Rights, the ideals of democracy, freedom, and the free and open society. It is our utmost responsibility and obligation to work collectively toward increasing educational standards, literacy, cultural enrichment, and gender equality, and decreasing racial discrimination.

Humanism is a philosophy, worldview, or life stance based on naturalism. Therefore, the meaning and purpose of life for humanists, in the context of science and philosophy, is to better understand them and work toward a comprehensive knowledge about the philosophy of life, the universe, the discoveries based on scientific methods, which we experience every day as they significantly affect virtually every aspect of our lives. Advances in scientific medicine, modern techniques of surgery, anesthesia, pharmacology, and biogenetic engineering have tremendously improved our prospects for a happier, healthier, longer, and more fulfilling life. Scientific research has and will continue to advance our knowledge of the universe and our place within it. So it is the meaning and purpose of life to pursue these ends. So coming back to the point stated earlier, the meaning and purpose of life as humans is not only to survive and replicate but also to derive pleasure and happiness while doing so. It is to love life in its abundance, to seek and work toward creating and instilling more abundance, and to share this love and experience with others.[177]

How can we help humanity by adopting humanism? It is the key question that needs to be answered. Let us see how much we can justify and answer from the discussion below.

Our world is in great trouble because of human behavior found on myths and customs that are causing the destruction of nature and climate change. We can now deduce the most simple science theory of reality—the wave structure of matter in space. By understanding how we and everything around us are interconnected in space, we can then find solutions to the fundamental problems of human knowledge in physics, philosophy, metaphysics, theology, education, health, evolution, ecology, politics, and society. This is the profound new way of thinking that Einstein realized that we exist as spatially extended structures of the universe—the discrete body as an illusion. Given the current censorship in physics/philosophy of science journals (based on the standard model of particle physics / big bang cosmology), the Internet is the best hope for getting new knowledge known to the world. But that depends on us, the people who care about science and society, to realize the importance of truth and reality.

Let us think this way: seven billion people on earth believe in one common thing—that is, humanity. But all do not belong to one religion. Also, the fact that humanity is the key theme of all religions means that humanity is the backbone or lifeblood of all religions. So humanity is the religion of all religions. Is it not better to be associated with the religion of religions—humanity—instead of associating with one component of humanity (that is, one specific religion)? Well, in fact, as stated earlier, about 1.5 billion people of the world do think this way, as they do not associate with any religion.

By associating with one universal religion (humanity), seven billion people can breed love among them. Humanity, with the quality of being humble, brings inner peace. We don't need a religion to have morals; if we can't determine right from wrong, then we look for empathy, not religion. Humanism is important because of the fact that having a nonsuperstitious worldview allows us to make more ethical choices based on the general desire to do the most possible good. In terms of the reality of the world today, all of the world's major religions, with their emphasis on love, compassion, patience, tolerance, and forgiveness that can promote good values, do not appear to be adequate in recent time. The grounding ethics in religion is not too good to appreciate. I hope you don't disagree with this. Please think rationally; think outside the religious box.

Looking at the fact of what function humanism serves in the society, it may be called another religion (in the sense of universal religion). It is a worldview that gives answer to the questions about human beings such as the place of humans in the universe, relations to one another and to other creatures on earth, and many more at the microlevel of human lives. Humanism is a basis for human values, ethics, and morality. There is another important difference between religion and humanism that is in the role of science. I would consider thinking that humanism is superior to religion in the sense that humanism accepts established outcomes of science. What is very important to look at in terms of values and ethics of life is that humanism is a progressive philosophy of life that, without religious dogmas and supernatural beliefs, our ability, human power, and responsibility to lead ethical lives of personal fulfillment and collective welfare, aspire to the greater good of humanity. It is the study and knowledge of what it means to be a good human being. It is not about dogma or absolute prescriptions for living our lives; rather, it is about striving to do good and better. Bertrand Russell said, "A good world needs knowledge, kindness, and courage. It does not need a regretful hankering after the pastor, a fettering of the free intelligence by the words uttered long ago by ignorance." Let us refresh our thinking by the following comparisons: reason, not superstitions; ethics, not dogma; respect, not worship; courage, not fear; morality, not religion; clarity, not delusion; good, not God; skeptic, not cynic; pragmatism, not ideology.

Humanism emphasizes human concerns as the basic principle. It overcomes tradition, religious dogma, or creed. Humanists seek to discover what best promotes human flourishing while leaving behind those religious beliefs and practices that would prevent humanity from achieving its full potential. This principle can be expressed in some core values such as reason, compassion, hope, etc. Humanists value reason and the use of the intellect and practices like the sciences and philosophy as the best way to generate and achieve accurate knowledge about the world we live in. They reject the supernatural, as promoted by religion, explanations for phenomena. According to Bertrand Russell, "Religion is something left over from the infancy of our intelligence." Similar ideas have been promoted by Richard Dawkins.

Humanists are driven by compassion, the idea that all people (regardless of nationality, ethnicity, race, creed, sexual identity, or other characteristics) are fundamentally of equal moral values. They also look to the future in the best hope with the belief that human beings, when they work together, can build a better world. It goes to suggest that it is the humanism rather than religions of varied ideologies that can and will make a better world with seven billion people. Humanism is a religion of mankind vis-à-vis a universal religion. Many tributaries of thoughts flow into the mainstream of humanist thought, but some are particularly significant. For example, much of modern humanism is inspired by the principles of Enlightenment, meaning that a commitment to reason is a tenet to change society and that a commitment to science is the best way to learn and understand the world and the laws by which the universe operates.

Humanists believe that people should be free to think and discuss any thought, including religious. Anyone, like humanists, has the right to

215

question any aspects of religions without any fear of persecution or threat of death. Who in the twenty-first century wants to go through this kind of fear prescribed by religion? The other major factor that has significant influence derives from liberal religious movements, including liberal Christian and Jewish movements such as transcendentalism and Unitarian Universalism. Such movements seemed to have significantly de-emphasized the role of God and the supernatural, thinking alike humanism.

It is important to note that the full range of values and ideals central to humanism is not so easy to capture in a short statement or through brief dialogue. The task is more challenging because humanism is nondogmatic by ideology, concept, and design. There are no required "creeds" or "religious dogma" in which humanists believe. There is no holy book of humanism that lays out what humanists should or should not do or should follow word for word like scriptures or holy books of religions. Like any tradition, humanism has no single founder and has no ultimate authority, but it promotes that ethics is an ever-changing field of human practice that must be altered to fit the context and the times.

However, humanists follow a creedal document, which is a set of *Humanist Manifesto*s, a record of consensus view of what humanists believe at a particular time, subject to be revised when circumstances change. Humanism is not inherently "antireligious" in the sense that humanism does not assert that all aspects of religious practice are harmful and inhumane; at the same time, humanism is not inherently "proreligion" because it does not claim that all elements of religious practice are positive and valuable. Rather, humanism eliminates aspects of religious practice that are found to be harmful, inhumane, and dehumanizing. Humanism is based on those elements that affirm and promote human flourishing, including moral, ethical, and all aspects of character development. The question is how humanists see religions. In response, humanists' optic is that they see religions and religious practices as human created and that they seek to ensure that religions that do exist need to serve human concerns rather than dictate them. Nevertheless, humanism lives and excels wherever human beings reach out to better understanding of the universe and our role within it, wherever human concerns are placed above the will of a god or the needs of a tradition or religion and wherever people believe that a better world is possible in *this* life. We, therefore, seek the values and importance of humanism to a wider reach among the human mankind.[178]

Is humanism important to the future of human beings? I think it is, and millions of others will agree with me. Humanism promotes that religion is not adequately designed to serve human concerns and, therefore, should not be allowed to dictate how societies should operate. All religions must operate by human spirit and within a framework of law determined by democratic mandate. Since the patriarchal religions depend on the ideologies so old that many of them have no relevance to contemporary conditions, they are unable to give any help in response to various problems of the world and its inhabitants. Indeed, fundamentalists adherent to these ideologies deny the validity of science. They are the most dangerous group of people on earth today. Since humanism and secularism are closely related, humanism is vital for our future. We need to change the way we do things so that science and a real understanding of the long-term consequences of what we do become the basis of our decision-making but not the incoherent perception of the Dark Age scriptures and holy books or the obsession that they tend to encourage in the short term.[179] On the contrary, humanism is a rational and intelligent approach to problem-solving that is geared toward making the world a better place for humans to live, is the only approach that will help us achieve our potential as a species, and gives us a chance at avoiding extinction.[180]

Humanism offers us one of the most important drivers of change that can improve our future. It is an optimistic view that humanism has an important future for human beings. So one may argue that it is possible to imagine a future without the tyranny of religious myth, dogma, and superstition. Perhaps we are not very far away to have the change. If our generations will not see, then the future generation will. It is important that while educating our children, we should encourage them to question everything, not be satisfied with blind faith and unsubstantiated claims but be skeptical of a priori beliefs, not only of their own but of their parents' or their teachers' as well, encouraging skeptical thinking and promoting a culture by which questions may be satisfactorily answered with logic, reason, and scientific evidence. This will help prepare them to be good and responsible citizens who can address the demands of a democratic and free society. Evidence suggests that seeds of religious doubt are planted more among the scientifically literate population, especially among the younger generation. It is always good to be skeptical, especially about ideologies you learn from scriptures and religious figures.

Studies suggest that when people, especially at a younger age, are being skeptical to anything doubtful and able to open questions, they could make people better lifelong learners. This, in turn, means that people who perceive their views on evidence rather than faith are likely to be better citizens. This kind of learning through education offers the best opportunity to help people be open thinkers and immunized against the intellectual virus associated with dogma and superstition. When people develop this mind-set, they would be able to publicly accept and promote the fact that many of the claims of the sacred books of the world's major religions are not valid.

We must understand the beauty of science—physics, mathematics, and cosmology. If we do so, only then can we question many texts written in the holy books—for example, the sun orbiting the earth, the universe being created in six days, Prophet Muhammad's night journey to meet God, the Virgin Mary being pregnant without having sexual intercourse with a man, and many others. We all know that there are places in the world where one risks decapitation for questioning certain religious claims. However, in a rational world, one can argue that questioning many religious claims, which are dubious in nature, should be viewed as inappropriate. Christopher Hitchens said that religion poisons everything. This can be debated, but with moderate view, one can argue that in the current environment, religion has devastating consequences in the political process in many countries in the world.[181]

Many Christians do not consider their faith as the sole source of wisdom in the world, although they may consider Christianity as the best religion. This understanding, to a great extent, allows the people to learn from other religions as well. The same is true for many Jews, Muslims, Buddhists, and Hindus. It is true that religion does not change, because it claims to be the eternal guide for solutions to the problems of mankind; as such, for anything that claims to be eternal, do not accept and agree to change. But the followers of religions do. Christianity has not changed, but Christians have changed. Hinduism has not changed, but many Hindus have changed. In the case of Islam, not only the fact that it has not changed, but also it neither was expected to change nor had any windows to change. This is the biggest problem in Islam, because Muslims refuse to change. Refusal to change (for any of the religions) makes religions the ultimate goal for all believers who want to maintain the status quo. In fact, religion becomes the instrument of holding on to power in the hands of orthodox rulers and their agents.

Prof. Lawrence Krauss asked, would it be naive to imagine that we can overcome centuries of religious intransigence in a single generation through gathering knowledge, education, and thinking outside the religious box? Perhaps not, but the stakes are very high not to try. As Feynman warned us, "It is our responsibility to leave the men of the future with a free hand. In the impetuous youth of humanity, we can make grave errors that can stunt our growth for a long time. This we will do if we, so young and ignorant, say we have the answers now, if we suppress all discussion, all criticism, saying, 'This is it, boys! Man is saved!' Thus, we can doom man for a long time to the chains of authority, confined to the limits of our present imagination."[182]

Humanists think that the entire field of human development is possible within the human frame. It is arguable that some form of religion is probably necessary. However, instead of worshipping supernatural figures and God, it is wise to represent the higher manifestations of human nature in art, love, science, rational thinking, and intellectual works. For many years, humanists and rationalists have worked to help people get rid of superstitious thinking, mean-minded self-interest and tried to bring the light of reason and empathy.

Humanism over religion—how we can justify? Let us talk about it and see if it makes sense. Humanism had struggled so long into existence within global societies by rejecting religious beliefs of all kinds. It emerged from societies where obedience to religious authority was imposed and strongly enforced in the society. It flourished within cultures where scientific evidence about the natural world was not only controlled but also suppressed or denied by the religious authorities when it did not fit the prevailing faith. For thousands of years, it has been witnessed that any skepticism of religion has been responded with hostility by the religious followers. As a result, many humanists have paid a heavy personal and social price. Despite the struggles, humanism has grown rapidly and has made much headway into the societies across the world. Most of the religions are more or less similar in believing supernatural components, but they are different in the way followers conceive, describe, and practice them. The imaginative impulse that has been perceived by the followers gives rise to thousands of different religious beliefs, which, in most cases, contradict one another.

To know why humanism is taking over religion, it is important to understand why religion came into being in human minds. Earlier discussion suggests that throughout human history, people have been asking questions about nature, the lives on earth, the universe, and so on. Zelda Bailey discussed briefly about humanism and religion during introductory talk to the newly formed East London Humanist Group in September 2012. His explanations seem logical and rational, at least to a humanist's perspective. We know that all creatures, including humans, need to survive and reproduce in the face of innumerable natural odds and, therefore, need to make sense of their natural environment, which is extremely complex and harsh. In a humanist's perspective, the whole idea related to those questions is central to making sense based on the tenet that there are no predetermined guidelines, dogmas, or doctrines derived from God or gods. So humanists have to construct convincing and satisfying values, principles, and ideals on their own in the absence of ready-made answers. The thoughts developed in this manner rely on personal responsibility, kindness, the wish to reduce or end fear but not cause suffering, and respect for the rights of others, as they are available by the most accurate knowledge of the world currently available. During this process, the rationalistic view of our nature drives us to question ourselves, and we are bound to answer satisfactorily and convincingly.

In a religionist's perspective, on the other hand, many explanations, rules, and pathways are provided ready-made in sacred texts and books, ancient rituals, and the declarations of those who claim special knowledge through revelation, such as prophets, saints, monks, and religious thinkers. Thus, it is the acceptance of religious beliefs that constitutes their faith. Religious people insist that these are religious questions and can be answered only by God or other supernatural forces. In a humanistic point of view, answers and explanations that rely on God or supernatural causes are irrational and unconvincing. But why are religious answers so widespread and deeply embedded in societies and cultures across the world? The most obvious answer is that they took root many thousands of years ago in a prescientific age when knowledge of the world was limited and information contained in oral accounts was passed on from one generation to another.

As we know, to survive, people deal with a hostile and unpredictable world, people created fables and stories to account for dangerous phenomena, and they devised imaginative ways for dealing with them. In the prehistoric time before writing was invented, oral storytelling was the primary means of transmitting historical knowledge and practical advice. Eventually, many of these ancient stories were written in the religious scriptures and holy books at a later date. Whenever the natural cause of something was not understood, a supernatural god, spirit, or force was imagined to be the actor of that cause. This interpretation appeared to provide certainty in a confusing and frightening world, obviously, without any scientific evidence. By accepting the truth of religious interpretations that could not be disproved and rejected, believers of all religious faith could be reassured, pacified, and comforted.

In societies where religious belief, practice, and social expectation out of those beliefs are very strong, it is extraordinarily difficult for an individual to take a contrary path. If someone does so, it is likely to have profound family and community repercussions; even the person is likely to be cast out from the family and community. Anyone who makes the decision to act in this way under these circumstances takes considerable personal courage and strong determination. One could imagine that the fear generated in this circumstance prevents someone from doing so. In other words, this fear acts as a powerful inhibitor. Yes, the power of religion is so strong that it shapes various customs by faith, which affected in the past and continues to affect societies now globally. However, many of these religious customs, such as discrimination based on caste, the ritual mutilation of children, the torture and murder of witches, etc., are intolerable and unacceptable in the world in the light of human rights, scientific knowledge, great advances in education, and mass communication and technology. Therefore, humanists embrace pluralism and strongly and vehemently oppose the harms and injustices carried out in the name of religious rituals, traditional rules, obligations, or customs. Humanists also confront the governments that allow them to continue. The voice of humanists is clear and loud—down with human rights violation, torture, and killing of people, including honor killing, by the name of religion.

Obviously, though, it is not possible, even with the scientific methods, to know the entirety of a disciple, because science is an ongoing process, and knowledge gained by science is based on results, which are provisional and subject to further research and analysis. Yet science does provide us with a sounder foundation for confidence and trust than any religion. For humanists, it is the explanatory power and integrity of science that make us understand the world and the nature around us. The fact is, where evidence is strong, faith diminishes. Although debatable, one can argue that humanism is taking over the place of religion in the prevailing modern, civilized societies across the world.

The religious path to morality follows a code of conduct, a set of commandments, or rules revealed by a god or other spiritual authority such as a prophet, a priest, saints, etc. The humanist, on the other hand, finds morality in the natural world through nonreligious ways. Humanists emphasize that the concept of morality has developed over thousands of years of social and cultural evolution revealing evidence to show that morality development is the natural continuation of the manner in which social relations have evolved in earlier species. As human beings, by our intelligence and superior conscience, we can modify values and morals by experiencing and practicing them. This is how we develop our morality. As these are natural phenomena, we have no need to look for supernatural explanations beyond nature to learn both moral and immoral behaviors and activities.

Each religion has its own peculiarities, but all religions lack universal applicability and acceptance, as they are unscientific and culturally dependent. Humanism has the ability to address the questions of life, the world, and nature in terms of realistic approach, which is common to everyone in every society. The ethical principles of humanism are shared values, unlike closed, religion-based values. Humanism is evergreen, always open to improvement as knowledge evolves and progresses in human society. Its secular entities go beyond religion and reach the global community with a common tenet (rather than separate and competitive manner by various religious beliefs and faiths; each proclaims its own ritualistic merits). This is the beauty of humanism, and this is why it is better than religion.

Last but not the least, humanism unites all people under one umbrella called humanity into one holistic community; it inspires global ideals and standards. It recognizes the fact that life on this planet is the only life and that its diversity is precious. In this regard, humanism considers humankind as just one species among many with no right to do harm to anyone and any species and the environment on which we all depend. Humanists emphasize that it is our world and that everything in it is final for us and that, therefore, it is our responsibility and obligation to pass on this planet to future generations in the best possible condition we can. There had been, and still are today, clashes and hostilities concerning the reality, the truth, and what these mean to human life. Clashes occur not only between religions and nonreligious groups but also within religions. It is essential, in the interest of the global humanity and for the future of our global society, that we find some common and unique values aimed at living peacefully in a harmonious way. To resolve this, humanism proposes that we should put aside religious differences, concentrate on our similarities, and strive for humanity. Once again, I loudly say, humanity is a better option over religion to lead a meaningful, adventurous, and happy life.

Transhumanism

Julian Huxley, who coined the word *transhumanism*, defined it as "Man remaining man, but transcending himself, by realizing new possibilities of and for his human nature." Transhumanism is an international intellectual and cultural movement supporting the use of new sciences and technologies to enhance human cognitive and physical abilities, which, in turn, find answers to the vital sufferings from undesirable and unnecessary human conditions, such as disease, aging, and death. As mentioned in the introduction chapter, it is a new philosophy that promotes the ideas of humanism in a new world where science and technology are the driving forces of change.

As humans, our future knowledge should be more rational so that it can not only promote traditional means of improving human nature (such as education and cultural refinement) but also direct application of science and technology to overcome some of our basic biological limits. This is happening now as we can see in the advancement in fields of science and technology. Technology has shown dominance over biological processes, which is gradually overtaking biology as the science of life. We can foresee the future that humans are not a finished product. They are evolving organisms, waiting for the right conditions to blossom. We can and we must evolve beyond natural and biological limits.

Our understanding about ourselves and about our relationships with nature around us has increased significantly. Since English naturalist Charles Darwin first published his *The Origin of Species* in 1859, it has become clear to the scientific community that species evolve according to interactions among them and with their environment. Humans are not the end of evolution but just the beginning of a better, conscious, and technological evolution. One of the core concepts in transhumanist thinking is life extension. Through genetic engineering, nanotech, cloning, and other emerging technologies, extended life will be possible. Transhumanists are interested in the ever-increasing number of technologies that can boost our physical, intellectual, and psychological capabilities beyond what humans are naturally capable of at present. Whether it is brain implants, bionic body parts, test tube babies, robotic hearts, exoskeleton suits, artificial intelligence, or gene therapies or genetic engineering that is likely to eliminate natural death, transhumanism is obviously an uncharted territory for the human species. Humans must be considered as a tenuous entity that is related to the animal and the "natural" as well. Humans are at a crossroads like other natural species that are reclassified in the face of new dynamics and shifting epistemological paradigms.

Homo sapiens is the first species in our planet that is conscious of its own evolution and limitations, and humans will eventually transcend these constraints to become enhanced humans, transhumans, and posthumans. It is conceived that the process may be rapid like caterpillars becoming butterflies, as opposed to the slow evolutionary stages from apes to humans. Future intelligent life-forms might not even resemble human beings. These posthumans will depend not only on carbon-based systems but also on silicon and other platforms that might be more convenient for different environments, like traveling in outer space.[183]

The general expectation is that, in the near future, greater manipulation of human nature will be possible because of the adoption of highly sophisticated techniques that are likely to be available, such as machine intelligence greater than that of contemporary humans, direct mind-computer interface, genetic engineering, nanotechnology, etc. Scientists have developed an implant that can translate motor neuron signals into a form that a computer can use, thus opening the door for advanced prosthetics capable of being manipulated like biological limbs and producing sensory information. This is on top of the earlier development of cochlear implants, which translate sound waves into nerve signals, which are often called bionic ears.[184] Transhumanists hope that by responsible use of science, technology, and other rational means, we shall eventually manage to become posthuman, with greater capacities compared to the present human beings.

Our nature as human beings has been changing by the advancement of technology. We are likely to see that in the next half century or so, the relationship between human and machine would be redefined in a profound way. Information sharing is a notable example as to how it works. This human activity is performed by billions of people in the world via digital medium, which did not exist a generation ago. Although the machine code is virtual and intangible, the information conveyed is real and meaningful. Christopher Phillips said, "Welcome to the future. You are now living in an era of transhuman technology—an era we may call an evolutionary renaissance. And you are what we call a cyborg: part human, part machine."[185]

We see that technology continues to advance. For example, research on brain- and body-alteration technologies has been accelerated under the sponsorship of the US Department of Defense. The department is interested in the battlefield advantages that they could provide to the super soldiers of the United States and its allies. There has already been a brain research program to "extend the ability to manage information," while military scientists are now looking at stretching the human capacity for combat to a maximum of 168 hours without sleep. Neuroscientist Anders Sandberg has been practicing on the method of scanning ultrathin sections of the brain. This method is being used to help better understand the architecture of the brain. As of now, this method is currently being used on mice. This is the first step toward uploading contents of the human brain, including memories and emotions, onto a computer.[186]

The book cites examples of as many as ten different technologies that can bring significant changes and transformations on human lives. They are the cryonics, virtual reality, gene therapy / RNA interference, space colonization, cybernetics, autonomous self-replicating robotics, molecular manufacturing, megascale engineering, and artificial general intelligence (AGI). There are certainly more on the way. Ours is an exciting time because science and technology have opened the doors that we only dreamed in the past. We envision the future, what's coming in twenty-five years, in fifty years, in one hundred years, or even the far future. Someone said, "Where there is no vision, the people perish." But predicting the future a century ahead or so is a difficult task. It is a challenge to dream about future discoveries and inventions through technological advancement that will alter our blind faith and beliefs and eventually establish humanity being liberated from the religious box. Science is the engine that already started changing the foundations of civilization and leading us to a new century with wandering inventions. Only in the last few decades, so many scientific discoveries have been accumulated that have not been achieved in all human history. Scientists foresee that by the next century, the scientific knowledge and discoveries will add many times over (Kaku 2011).Today, we do not think lightning bolts, tsunamis, earthquakes, plagues, and diseases are the act of God or gods. Our ancient ancestors did not see the magnificent achievements of science and technology. For example, jet planes fly much higher up beyond clouds, rockets explore the moon and planets, MRI scanner peers inside the living body, cell phone keeps us talking with anyone in the planet, and so many. Science is dynamic, and it is expanding exponentially. Scientific discoveries and innovations are changing every aspect of human lives and social landscapes, which, in turn, is going to alter old traditions, beliefs, and prejudices. Scientists believe that one hundred years from now, we will be able to manipulate objects with the power of our minds rather than by the power of God or gods. Here are some examples: Computers will be able to accomplish our wishes, and we will be able to move objects by our thoughts (a telekinetic power that now belongs to God or gods). With the power of biotechnology, we will be able to create perfect bodies and extend our life spans. With the power of nanotechnology, we will be able to take an object and transform it into something else; we will be able to harness the unlimited energy of the stars. While these are unimaginably advanced powers of mankind, which seem to be godlike power, the works of scientists all over the world have already started working in respective fields. The equations of the quantum theory are the basis of the creation of universe, and the culmination of all upheavals is the formation of a planetary civilization, what physicists call a type 1 civilization. This civilization becomes the greatest transition in history, making a significant turn from all civilizations of the past, which have impacted culture, religion, business, trade and commerce, entertainment and other activities, and even war. Scientists believe (based on the calculation of the energy output of the planet) that we will attain type 1 civilization within one hundred years from now (ibid.).

Science has given us a lot; it has taken us from the depths of darkness to the threshold of the stars. Science is unbiased, fair, and neutral. Science is a double-edged sword (mentioned in chapter 3, page 153), but how this powerful weapon can be used depends on the wisdom of users. Einstein said, "Science can only determine what is, but not what shall be, and beyond its realm, value judgments remain indispensible." We witnessed the massive destructive side of science during WW I and WW II and the killing of hundreds of thousands of people by poison gas, automatic firearms and machine guns, and the atomic bomb. Nonetheless, science helped rebuild humanity above the ruins of wars and made peace and prosperity for billions of people. It suggests that the true power of science enables mankind to do better. Science helps us innovate, create, and endure spirit of humanity, and at the same time, it helps minimize our deficiencies, because we can organize knowledge in a systematic manner (ibid.).

Conclusion

Religion books talk about good and evil, heaven and hell, salvation, morality, God's law, and holy books that are the words of God, which were revealed to God's chosen people, the son of God, and the prophets and avatars. It sounds interesting indeed. Who does not want to go to heaven where seventy beautiful *hurs* would be allocated to a single man for pleasure? Who wants to be burned in hell? Today, seven billion people on earth deserve the same air to breathe, enough food to eat, and the same treatment of God. There are thousands of questions one can ask and seek answers to. What is the truth? There should be the truth that is the same for all of us in this planet. Believers of one religion claim that their god's words are true, which differ from other believers of different religions. The basic question is, what is the truth? According to Schopenhauer, "A religion has never yet, either directly, either as dogma or parable, contains a truth" (cited in Friedrich Nietzsche's *Human, All too Human*, 1986). Those who validate science find no truth in religious claims. One can strongly doubt that thunder is caused by the anger of Zeus or that Allah is the one true god. Many see them as myths to explain the unknown, to give some kind of meaning within a culture or religion, or to empower the ruler. They were all human imaginations.

We, the humans, are a curious species. We wonder and seek answers. We want to understand the world, the nature of reality, to know where we come from, who created the universe, and so many other questions. According to scientific determinism, the laws of nature tell us how the universe behaves, but they don't answer why. Most people say that God chose to create the universe that way. Stephen Hawking (2010) noted that if the answer is God, then who created God? That means some entity exists that needs no creator. This is known as the first-cause argument for the existence of God. He, however, claimed that it is possible to answer these questions within the realm of science without invoking any divine beings. Let us agree that God is the first cause, which requires no creator. But the god and religion are not the same. Religion is a de facto organized system of belief. One can believe in a god without belonging to any religion or religious organization. Religion is nothing more than one of the ways, among others, people worship God, but it may also be a barrier between man and God. Abandoning organized religion may free a person to find the true divine god for what he really is. There are many religions but only one god. You are free to believe in a divine force or god without adhering to any religious dogma or doing an outward demonstration and rituals, such as worship services, social functions, religious traditions, etc. Your beliefs may not fit neatly into any religion anyway, and there may not be any group or ideology that accurately represents your viewpoint.

So what is the stake? We all are humans on this planet regardless of our religions, castes, colors, or ethnicity. But we can all be the light in someone else's life. We humans made religions but not the other way around. What we must learn is that the search for God is nothing but a search for the good virtues in ourselves and in others. I think that the biggest religion in the world should be humanity with over seven billion followers and is ever increasing. If we believe in spirituality, we believe in humanity. "There is but one law for all, namely that law which governs all law, the law of our Creator, the law of humanity, justice, equity, the law of nature and the law of nations," said Edmund Burke.

It is intellectually very difficult to step outside the established wisdom of our time and place, and to openly stand against it is even more difficult. It is conceivable to think how our individual conviction of what is true or right shows fear before the pressure of our peers. This is not specifically the Western tradition of courage in the advancement of free speech and freethinking. Rather, it can be traced from ancient Athens to modern Western countries. It would also be quite wrong to think that this norm of freethinking is confined to the West. Consider, for instance, the Chinese dissident Liu Xiaobo. Liu was sentenced to eleven years' imprisonment in 2009 for free speech against state power. In his book *No Enemies, No Hatred* (2012), he quotes a traditional twenty-four-character Chinese injunction: "Say all you know, in every detail; a speaker is blameless, because listeners can think; if the words are true, make your corrections; if they are not just take note." During the Arab Spring of 2011, a "day of rage" was proclaimed by dissidents in Saudi Arabia; he is Khaled al-Johani. He was condemned to eighteen months' imprisonment for demanding the right to speak freely.[187] These are only few examples. There are innumerable cases, especially in the countries ruled by dictators where democracy is nonexistent. This kind of situation is more rigorous in the case of religion; believers of religion and religious dissidents of particular religions are threatened of their lives if they step outside the religion or are skeptical about the holy texts. A religion proclaims that there is one and only one book that will reveal the word of God. But humanists argue that the holy books are not the message for mankind sent down by God. Rather, they are mundane documents. Texts are the writings of people at different points over a long stretch of time and in different historical, social, and political circumstances that were passed down through generations.

Man is born free and free to work out his own destiny. He can take the responsibility of his day-to-day life and so the opportunity. This sounds like humanity, which originates from freethinking and takes everything based on scientific inquiries and validations. If one puts the argument that when religions are reformed and made rational, then religions are good. So this requires reconciliation of science and religion to a certain standard. Is it possible? It may or may not. But Bertrand Russell has given a reasonable answer to this question and said, "The answer turns upon what is meant by 'religion.' If it means merely a system of ethics, it can be reconciled with science. If it means a system of dogma, regarded as unquestionably true, it is incompatible with the scientific spirit, which refuses to accept matters of fact without evidence, and also holds that complete certainty is hardly ever impossible." He further said that the organized religions that have dominated large populations have involved a greater or less amount of dogma, except Buddhism, especially in its earliest forms.

We all are alike. We have senses that perceive and react; we have knowledge to analyze and integrate. Our only difference in this regard is that some have judged (or assumed) the premises of their religions to be factual, while others have not. Is it not the fact that one would not want to commit to an idea of which one is not convinced? I welcome you to try to convince me that your religion is true, but I am not going to just believe by blind faith. I may ask why there are so many errors and contradictions in religious scriptures. Truth is not determined by vote. Rather, it is determined by proof. What could be said about a religion that, if true, would make it false? If you can't answer that question, then your conclusions will not be based on honesty. You can't expect me to respectfully listen to you if you are unwilling to accept that you might be wrong. I am open-minded and willing to change my position, if needed. Can you also be fair enough to follow the facts wherever they lead? Many unbelievers have carefully considered these questions, perhaps even more deeply than you (believers) have. Humanists were, at one time, just as religious as you are now. At some point in their lives, they found religion irrelevant and moved away from unfounded blind faith on religion.

After an honest examination, I am convinced that the holy books are written by people and that the texts are based on stories, myths, and dogmas prevalent in a specific time period where there is no evidence to support them. In an open forum, the arguments may be strong to sufficiently justify the fact that religions are not more moral or tolerant than humanists, agnostics, and atheists and that religion has caused more harm than good. Why are our conclusions not more valid than yours? I deeply sense that there is no need to worship, confess, or apologize to any supernatural power as religions suggest. As freethinkers, we get along with the scientific method and the rule of reason to understand the natural world, instead of believing in the existence of sky god who can't be reached by such methods or rules. This is what humanism is; it is simple to understand and relates to science, reason, and progress of mankind. Isaac Asimov, a scientist and a humanist, emphasized humanity to be superior to religion and stated that it is humanity that alleviates human and social ills with the help of science. Others have the similar views on religion, God, and humanity. Albert Einstein said, "I am a deeply religious nonbeliever. This is a somewhat new kind of religion. I have never imputed to nature a purpose or a goal or anything that could be understood as anthropomorphic. What I see in nature is a magnificent structure that we can comprehend only very imperfectly and that must fill a thinking person with a feeling of humility. This is genuinely religious feeling that has nothing to do with mysticism. The idea of a personal god is quite alien to me and seems even naive."

"Great men think alike" is a saying I learned (you, too, I suppose) as a teenager. Who hasn't heard the name Stephen Hawking? Commenting on God and science, he said, "The question is, is the way the universe began chosen by God for reasons we can't understand, or was it determined by a law of science? I believe the second. If you like, you can call the law of science God, but it would not be a personal god that you could meet, ask questions." Distinguished people like Albert Einstein, Stephen Hawking, Isaac Newton, Galileo Galilei, Blaise Pascal, Johannes Kepler, Charles Darwin, Michael Faraday, Ernest Rutherford, Erwin Schrodinger, Max Planck, Carl Sagan, Julian Huxley, Edwin Hubble, Max Born, Francis Crick, J. Robert Oppenheimer, Archimedes, Winston Churchill, Bertrand Russell, Jean-Paul Sartre, Friedrich Nietzsche, Voltaire, David Hume, Bill Gates, Steve Jobs, Warren Buffett, and you name thousands more with whom you can resonate to the voice of human civilization and scientific development mankind has achieved—we all owe it to them.

I think religion should be a private matter. I recognize that for many people, religion brings comfort, inspiration, and strength. I think people are free to believe whatever floats in their brain as long as they do nothing to hurt others and stop to think freely and question even the religion per se and, thus, stand on the path to human progress. As I understand, religion has no place in public life. In my teens, I listened to people about religion but did not heed much attention. As I became mature, I started thinking about the religions somewhat superficially. When I reached a much higher level of understanding, I learned that religion has nothing to do with logic, reason, or rational worldviews. I was not convinced about the rules and rituals of religions. I realized that knowledge comes from the experience of the material world rather than hypothetical paradigm or blind faith. Mao Tse-Tung said the following: *If you want knowledge you must take part in the practice of changing reality. If you want to know the structure and properties of atom, you must make physical and chemical experiments to change the state of atom. If you want to know the theory and methods of revolution, you must take part in revolution. All genuine knowledge originates in direct experience.*

We often ask ourselves how and from where we get religious faith. There are several arguments that prevail, such as religious faiths that come from upbringing and shared cultural heritage and cultural condition (Loftus 2006). Philosopher of religion John Hick emphasized that it is the accident of birth that determines religious affiliation. For example, someone born to Buddhist parents will be Buddhist, someone born to Muslim parents will be a Muslim, and someone born to Christian parents will be a Christian (John Hick, 1989, cited in *Why I Became an Atheist*). Richard Dawkins reiterated the same, more emphatically, that the majority choose the religion that their parents belong to rather than the sect, the miracles, the moral code, the best cathedral, and so on. Heredity is the key factor that matters. With the full knowledge of the arbitrary nature of this heredity, people believe the religion often with a fundamental ideology that they are ready to murder people who follow a different religion (Richard Dawkins, 1994, cited in *Why I Became an Atheist*). However, when some people claim that they do not share the same belief, they go out of sociological and cultural facts. They are above the so-called ideology. That is why Voltaire said, "Every man is a creature of the age in which he lives, and few are able to raise themselves above the ideas of the time."

Who would not agree that good people do good things in life and bad people can do evil with or without religion? But for good people (as they are religious) to do evil, that takes from religion, someone said. We would be rather better off without blind allegiance to the commands of religious prophets, preachers, bishops, and saints who claim those commands to be the absolute knowledge of God. Would it not be better to focus on how to improve our life on this earth rather than to focus on heavenly afterlife rewards that we have no knowledge about that fact? The irony is that all religions tell us that nonreligious people deliberately commit murder, rape, theft, suicide, as they deny that there is a god who watches over us and ultimately punishes us for our actions. This is not true, though. We must look at the evidence; many people who have rejected religion lead a happy, good, and productive life. Would it not be better to enjoy in alleviating someone's suffering for the pure pleasure of doing it rather than doing it for the false hope to be rewarded by God? Would it not be better to spend money on people's basic material needs rather than to build cathedrals, mosques, and temples for praying to God for false eternal rewards?

Elizabeth Anderson commented that the god is cruel and unjust and commands and permits us to be cruel and unjust to others. She also said that since we know that such cruel acts are morally wrong, we cannot take this literally, at its face value, as the evidence for theism in scriptures. She suggested that if someone hears a voice or some proof that purportedly reveals God's word that tells someone to do things that are wrong, then one should not believe it. I agree with her. Being rational and perfect and with a healthy mind, who would think that a perfectly good god would order to do wrong things? Not me! If I hear such a voice, I would think I am having a brain malfunction or having a bad dream. Raymond Bradley has argued from an atheist's point of view that if there are objective moral truths, then God does not exist. According to him, the objective moral truths are the following:

- It is morally wrong to deliberately and mercilessly slaughter men, women, and children who are innocent of any serious wrongdoing.
- It is morally wrong to provide one's troops with young women captives with the prospect of using them as sex slaves.
- It is morally wrong to make people cannibalize their friends and family.
- It is morally wrong to practice human sacrifice by burning or otherwise.
- It is morally wrong to torture people endlessly for their beliefs.

Would it not be reasonable to think about morality? These moral principles are objective indeed. All religious scholars admit these moral principles. Since these morally wrong things are permitted by a god of major religions, then that god does not exist (Loftus 2012).

Humanist moral and ethical systems of thought do not need a prior belief in God or any established religion, and morality does not intact within religion. Rather, it resides in the human mind. We inherit morality and ethics from our Paleolithic ancestors, and later we have refined them according to our cultural preferences and apply them within our own unique historical circumstance, said Michael Shermer. Abundance of evidence is available on humanist ethic. The basic tenet of humanist ethic evolves from human needs, such as physiological needs, safety needs, and physical needs, and, above all, we have a need for love and self-esteem. In all these needs, it is obvious that we have to be good to others if we want to fulfill our own needs. We need others to achieve things we want; we can't do it alone. This is especially true in the case of love that bonds people together. Clearly, we want the needs of the people we love to be met. This means doing good for the people we care about, which is extended to the community level at large. So rational people want their needs to be met only by doing "good" to others. This is human ethics. To become a good human being and to acquire human ethics and morals, one does not need to be a believer of a particular religion. If you ask believers what their prime identity is, you will be utterly surprised by their answers—"I am a Muslim" or "I am a professor" or "I am a Canadian" or I am a Hindu" or "I am a father" or "I am a Jew" or "I am a doctor" or "I am a Christian" or "I am a politician." What do you think? I think the answer is "human being."

Phil Zuckerman (2014), in his book *Living the Secular Life*, noted that religion is not the sole source, arbiter, or purveyor of morality and values. We have to look at this in the historical, sociological, and philosophical perspective. Based on secular ideologies, many of the greatest moral and ethical advances have been championed by the nonreligious people—establishment of democracy, women's rights movement, the fight against caste in India, creation of the Universal Declaration of Human Rights, to name a few. Sociologically, as I have highlighted in this book, people in many countries in the world with the highest proportion of nonreligious population tend to be more humane, having high morals and ethical values, compared to the societies that are most religious. Philosophically, secular morality is neither dependent upon religious faith nor linked to heavenly reward or punishment. The Golden Rule is the perfect example, which requires no boarding on faith. Because of our neurological capacity for empathy and biological evolutionary process, we have adopted the Golden Rule for thousands of years, Zuckerman pointed out. Humanists don't pray to god(s) for help and don't appeal to avatars or saviors or prophets to save them from adversities. They do their works themselves. This is very important, and humanists know it.

But only saying "I am a humanist" is not enough; what is important is that humanists must be known to him and to others by their actions. To this effect, working together for the greater good of all human beings is vital. Just think how fortunate we are as human beings having the attributes of creativity, higher intelligence, compassion, love, morals and ethics, and law-abiding quality. We, as human beings, are the torchbearers of civilization, and we have the power to build a global society where science, reason, progress, and humanity can grow side by side. With our amazing human quality, we can work together with integrity to make a better world where humanity thrives, and this, in turn, will force us to be good for the greater good without believing in religious dogmas, persecutions, hatred, divisiveness, violence, and so on.

What can we add to humanity? A good question indeed. Are we not incorporating science and technology into our daily lives? Yes, we are. Technology has enabled us to improve our lives in many ways. The way we produce art, science, literature; the way we communicate and educate; our sharing of ideas and experiences; even our interactions with our environment—all this has been redefined by the integration of technology into our society. Combining humanity with technology will help us all become more human.[188] We can see an exciting future. We, as humans, will apply this technology responsibly. New technology will redefine society, a society in which every human being will have the opportunity to fulfill their potential, regardless of their physical or mental condition. Although there are certain risks and some are likely to misuse any technology for personal gain, being aware of risks, as part of responsibility, will help society minimize serious dangers. Above all, I think that by accepting the reality (that is, integrating technological advancement into our lives) and adopting them, we can be better humans.

Billions of people are believers, in some way or another, of different religions, and they find comfort in the face of pain and suffering. There is nothing wrong in it. But when faith outweighs rational thinking, reasoning, and empirical evidence, here comes the wrong, because they don't accept the facts. Rather, they find various ways to reject them. There are many laypeople and scholars alike who feel that the orthodox religious believers could never live outside the context of the ancient or of the seventh-century origins. When applied to twenty-first-century science, logic, or humanistic reasoning, to it they fall apart. This is why the organized religions have always relied so heavily on the fear of punishment and rewards. For example, question religion, say anything malign to it, or leave the religion, and the consequence of it is that you may be killed. This is more prevalent in Islam. It is a totalitarian modus operandi that silences the dissents. There are problems of different kinds not only in Islam but also in other religions. Whatever the reasons may be, it is observed that more and more people prefer to live their lives without religion. The fact is, those who don't follow religious rules or practice rituals or live religion completely are not bad or immoral people. Rather, they are better humans than their religious counterparts. These people value reason over faith, prefer to face challenges by action over prayer, prefer to be freethinkers over obedience to supernatural authority, and prefer hope in humanity over hope in a god or deity. All nonbelievers do not completely rely on science and reason because of the fact that those are not sufficient, although scientific facts are necessary to rely upon. At the same time, freethinkers or humanists distrust anything that contradicts science. They might have different views on many things, but they strongly emphasize the free inquiry, open-mindedness, and unending questioning. One may argue that great literature, art, music, and many creative works of people in dealing with ethical and moral values and consciousness are better described than the mythical tales written in the religious holy books.

We are in the initial stage of humanist movement, which is only the beginning of the process to change our world. When 1.5 billion people worldwide are humanists, it is definitely good news that we are progressing well toward establishing the fact that humanism is superior to religion and is the best alternative of religion for a better life. Well, we may not finish the work, but neither can we just stop doing nothing. Let's do whatever we can to make a difference. Therefore, one of the objectives in writing this book is to explore and invite people to join the humanist movement and to show a glimpse of hope to the people at large, irrespective of their race, color, gender, religion, nationality, as to how we raise our children with love, affection, endurance, optimism, nurturing of independent thinking beyond religious orthodoxy; how we can face challenges of life without fear; and how we can build worldly morality based on empathy, the Golden Rule. What is the purpose of life as a humanist? The purpose of life is to live happily. What's in your head?

Religious people say we can get God through their religion (of course, only with which they are affiliated). This is not the whole truth. The right path to God is the spirituality, not the religion. Christopher Hitchens (2007), who was a humanist, rightly said that those who search for the truth or God do not need any priests, imams, rabbis, or any hierarchy above them to monitor or supervise their way of life, belief, and doctrine. Humanists seek truth through humanity. For them, there is no special place on earth holier than another; there is no place for pilgrimage or sacred wall or cave or shrine or rock. Someone said, "The essence of religion is to fear God and obey God, whereas the quintessence of spirituality is to love God and become another god." This carries profound meaning. Humanists think that religion is nothing but the adherence to a certain dogma or belief system. Spirituality, on the other hand, places little importance on beliefs but is concerned with growing into and experiencing the consciousness. Religion is a sweet escape—doing things in the name of religion even simpler. Religions are not free from fanaticism, and fanaticism is bad in all counts. Someone said, "All fanaticism is false because it is a contradiction of the very nature of God and of truth. Truth cannot be shut up in a single book, Bible or Veda or Koran, or in a single religion." So the best and only thing to think is "One world, one life, one god, one religion—humanity."

The Last Word

I tend to think that there is sincere intention to synchronize science, philosophy, and religion and build a common platform where we all can build a good and meaningful life together. If it happens, the world's seven billion people would be a boon to human civilization. But the relationship among science, philosophy, and religion is in turmoil. There could be reconciliation and would be possible in building a unified model for a better and meaningful life for mankind rather than living with faith clash among peoples. Let me hope that we are, as a global society, close to making that model. As Einstein, while replying letter to someone, said, "We are quite close to each other in essential things, that is, in our evaluations of human behavior. What separates us are only intellectual props and rationalism." But I think that science, philosophy, and religion, if used in isolation of humanity, will be dangerous. The world is wild with the delirium of hatred, and the conflicts are cruel. The age-old clash between cultures between religions is still causing pains and sufferings at the present time. We have to stop to happening it, we have to stop hating one another, we have to stop destroying civilization, and we have to stop killing humanity.

People are dynamic and are used to adapting new things. So there is no reason to believe that we will not change ourselves for the sake of the common good—a meaningful, enjoyable, and happy life. This goal can only be achieved through humanity, a unified model to lead life for global population. To be a humanist, you don't have to cut out the faith of God, but you have to conceptualize God in a rational way. God does not like to see us building animosity, fighting, or killing among people. If we stand by our conscience and exhibit truth in our words, deeds, and actions, we will stand by God regardless of our faith. But a religious name or title does not reflect the truth or make anyone superior to another. God and religious title should not be mingled up. God is omnipotent and omnipresent; he can be found all over. God does not reside in the church, in the mosque, in the temple, in the synagogue, or in any designated places of worship. Nobel laureate Rabindranath Tagore wrote the following

Whom do you worship in this lonely dark corner of a temple with doors all shut? Open your eyes and see your God is not before you! He is there where the tiller is tilling the hard ground and where the path maker is breaking stones. He is with them in sun and in shower, and his garment is covered with dust.

Who is not aware of the fact that the prophets were not especially learned individuals? They did not have a high level of education or intellectual sophistication. They certainly were not philosophers or physicists or astronomers. There are no truths about nature or the cosmos to be found in their writings (for example, the sun revolving around the earth is not true). God created the universe, so God knows the truth. What religions propagate is that every occurrence is the act of God. When God is omniscient and omnipotent, then not only every occurrence but also every human thought and action is his work. How is it then possible to think of holding us responsible for our deeds (good or bad) before such an almighty god? Is it not the case that in giving punishments and rewards, God would be passing judgment on himself? How can this be combined with the goodness and righteousness ascribed to him? said Albert Einstein. Religion is a spiritual propaganda. We should rather think this way—our lives should have been governed by longing for love, searching for knowledge, searching for truth, and sharing the sufferings of people and caring for them.

There are good words in the holy books, such as practicing justice and loving kindness to your fellow human beings. As Spinoza argued, scriptures are not the only work that is divine. If they are divine, then many scholars' works of literature are divine too. For example, if William Shakespeare's *The Tempest* or Mark Twain's *Adventures of Huckleberry Finn* is the words of justice and mercy or if Charles Dickens's *Hard Times* inspires one toward love and charity, then these works, too, are divine and sacred. The word of God, Spinoza said, is not confined within the compass of a set number of books. True religion requires no belief in any historical events, supernatural incidents, or metaphysical doctrines. The divine law directs us only on how to behave with justice and charity toward other human beings. What we need is to uphold justice, help the helpless, do no murder, covet no man's goods, and so on. All the other rituals or ceremonies prescribed in the holy books' empty practices do not contribute to blessedness and virtue. While discussing about Spinoza's thoughts about religion, Steven Nadler explained that all opinions, including religious opinions, must be absolutely free and open to express the truth. It is one's right. For conscious and literate people, it is impossible for them to be completely under another's control with respect to religious belief. They are not going to accept or endorse others' views because they are capable of making their own judgment on any matters whatsoever and are not compelled to accept others' views and opinions.[189] It is their natural right to choose their own views on any matters including religious faith. This is humanists' principle—better way of life over religious binding and a title.

Humanism rejects dependence on religious faith, texts in the holy books, resurrection, reincarnation, or anything that has no evidence whatsoever. Humanists take control of themselves with responsibility to lead an ethical, moral, and dignified life of an individual and for all others' fulfillment, wishing, aspiring, and working for the greater good of mankind and all other lives on earth. We don't have to fear the unknown and unexplained metaphor. Rather, we should hold courage from the wondrous discoveries to live with them and find hope in knowing life. When more than one billion humanists in the world feel that humanity brings better values, morals, and dignity than religious superstition, by living and thinking outside the box called religion, then why can't you feel the same way? People help people; this is what humanism teaches us. In reality, human guidance is the guidance we follow through in our daily lives.

Humanity stands for values that eliminate fear and solidify love, honesty, equality, kindness, compassion, and the attitude of treating people the way you want to be treated. Humanity is like the open sky where a few clouds cannot block the entire blue sky, and when the cloud gets thicker, it turns into rain that gives life. Some logics are very meaningful on their own merits, such as *reason, not superstition; ethics, not dogmas; respect, not worship; courage, not fear; morality, not religion; clarity, not delusion; good, not God*. It sounds humanistic. They make good sense to me. What about you?

We can understand one another (nonbelievers and believers) quite well for concrete things in life. We have the power to make life free and beautiful, to make life a wonderful adventure and memorable experience. Where our mind is without fear, our knowledge is free there. Let us work for a world of reason, a world where science, technology, progress, civilization, and humanity grow together hand in hand. Ignorance is the product of believing, whereas knowledge is the product of thinking. Freethinking is the first essential step toward achieving this worldview. Let us start practicing freethinking without the fear of punishment and threat of life.

Richard Feynman said, *"Religion is a culture of faith. Science is a culture of doubt."*

The essence of science is that it accepts and admires questions and doubts; it admits when it is wrong, and when challenged, science replies with evidence. On the contrary, religion teaches us not to question or doubt; it never admits when it is wrong (especially for the words of scriptures), and when challenged, religious believers become hostile. I finish writing the last word by saying this: God is for all, and so is humanity, but religions are not; set religions free, and a new religion will emerge, and that is humanity.

Purpose of Life

Purpose of life is to live.
All the efforts are nothing but to survive
Along with the history of time,
By carrying the ancestral genes to pass on to next generation;
The play of the survival game dictates to
Struggle for existence, power, wisdom,
Gaining edge and win over the competitors.

Earth's magnificent look at the moonlit night
Delivers the message of love;
Youthful earth on a freshly dewfall morning
Resonates with song lyrics and verses of poems.
Awakening earth in the late afternoon
Turns into gray lines of prose,
And that life struggles for another day.

Into the bones and blood, life shrinks for all,
Yet many survive long enough to enjoy life.
Even the naked, hungry, sick children want to live,
Though know death will happen soon to them.
Our primitive deeds, struggles, beliefs, and dreams
All but to live and in principal terms;
The purpose of life is to live;
All the efforts are nothing but to survive
Along with the history of time

Evolution of cosmic objects, galaxies, and stars,
Infiniteness of space and time and expansion of cosmos,
Incomprehensible universe from beginning to present to future,
Not known in definite terms but life evolved,
The identity crisis ingrained in human mind;
Being a part of earth at events of time,
We learned at least the future of life,
And being lived in the closed loop of life and death,
There comes to know the certainty of our destination.

The supernature's cosmological architecture tells,
Our destination has no future at the end,
But only know the purpose of life is to live.
All the efforts are nothing but to survive
Along with the history of time.
We, the humans, make our own purpose to live life
In the way we want.

—Abdur Rahim

Annexure 1

Major Religions*

Hinduism

It is one of the oldest living religions dating back to at least 2000 BCE. Hindus refer to Hinduism as *Sanatana Dharma*, meaning "eternal teaching" or "eternal law." There are over one billion Hindus in the world today. Hinduism comes from the ancient civilization known as the Indus Valley civilization, which flourished between 3500 and 1500 BCE. It is thought that this civilization was reformed when nomadic people called Aryans arrived in India and that Hinduism developed from the religious ideas of both Aryans and Indus Valley peoples. These Aryans had a collection of hymns about creation known as Vedas. The Vedas (which comes from Sanskrit, meaning "knowledge") became wisdom literature, which was considered an infallible source of timeless, revealed truth. The most important of the Vedas was the Rig Veda, which consisted of hymns or devotional incantations of more than ten thousand written lines in ten books.

On the metaphysics and philosophy of Hinduism beliefs and Hindu gods, all is one (Brahman). The metaphysical foundation of Hinduism, which is expressed in both the Vedas and the Upanishads, is that reality (Brahman) is one or absolute, changeless, perfect, and eternal. The ordinary human world of many discrete (finite) things (which our mind represents

* Source: *The Usborne Internet-Linked Encyclopedia of World Religions*, Susan Meredith and Clare Hickman; edited by Kristeen Rogers. A paper published internet-linked book. Catalogue: www.usborne.ca/ottawa.

by our senses) is an illusion. Through meditation and purity of mind, one can experience their true self, which is Brahman, God, the one infinite, eternal thing that causes and connects many things. True enlightenment is self-realization—to experience the supreme reality as self.

Hinduism has thousands of gods and goddesses. Each has his or her own special attributes and characteristics. Hindus usually worship the gods and goddesses of their choice. Most of the Hindus believe that all gods and goddesses, although of different natures, are of the same ultimate reality— the supreme Brahman. There are three main Hindu gods, *Brahma*, *Vishnu*, and *Shiva*, who are related with the creation, preservation, and destruction of the world. This cycle is thought to happen eternally, which has no beginning or end. *Brahma* (not the Brahman) is the creator, Vishnu is the preserver, and Shiva is the destroyer and liberator. Together, these three gods are known as the *Trinity* or *Trimurti*, meaning "three forms."

Vishnu, the preserver of the universe, is believed to have come down into the human world in various physical forms known as *avatars*. As the seventh and eighth *avatars* (Rama and Krishna), Vishnu walked on earth in human form. Rama and his wife, Sita, were known for their purity of character, love for each other, and other high moral values. It is believed that Sita was kidnapped by Ravana, the demon king of Sri Lanka. Rama defeated Ravana in a war with the help of monkey god Hanuman and rescued Sita. This victory is considered by Hindus as the triumph of good against evil. Krishna is the most popular of the avatars of Vishnu. Worshippers behold various aspects of Krishna's personality. While some look on his innocence and charm as a child, others look on his many pranks as a young cowherd and his pure love for a milkmaid, Radha.

Shiva (the destroyer or liberator) is shown with four arms, matted hair, a crescent moon on his head, and a snake twined around his neck. His upper right hand is shown holding a drum on which he beats out the rhythm of his dance—the dance of liberation and recreation. Shiva has a third eye in the forehead that symbolizes many things, including his wisdom. Three goddesses are associated with Shiva (Parvati, Durga, and Kali). Parvati, the beautiful and gentle wife, represents the compassionate side of Shiva's nature. Durga and Kali are fierce and powerful, who match Shiva's destructive side of his nature. Durga slays demons with a sword, and Kali is known as the destroyer of evil. Ganesha, a son of Shiva and Parvati, is the remover of obstacles. This is why Hindus worship him at the beginning of a new undertaking, such as the starting of a journey, at a wedding, etc. It is said that Shiva beheaded Ganesha in a fit of anger. Later, Shiva restored Ganesha to life by giving him an elephant's head.

The Hindu scriptures (such as the *Rig Vedas*, the *Mahabharata*, and the *Ramayana*) contain guidelines that inclusively define social status of Hindu people to work in harmony in the society. These groups are known as *Varnas*. The first *Varna* consists of priests and teachers known as *Brahmins*; they are the superior class of Hindus. The second in rank in status is known as *Kshatriyas*; they are the rulers and soldiers. The third in rank is known as *Vaishyas*; they are the merchants and farmers. Members of the fourth in rank are known as *Sudras;* they are manual workers. Apart from these four main Varnas, there are thousands of groups in the Hindu society called *castes*, or *jatis*. A person is born into a caste. Although the differences among castes are not as rigid today as they were in the past, caste is still important in defining a Hindu's place in the society.

According to scriptures, Hindus have four stages of life called *ashramas*. These are the student, the family man, the recluse, and the wandering holy man who cuts all family ties, owns nothing, and lives by begging. The ultimate spiritual goal for a Hindu is to achieve *moksha* and, thus, to be united with *Brahman*. Many Hindus go to pilgrimages. Each year, millions of Hindus make pilgrimages called *yatras* to holy places. The most sacred of all Hindu pilgrimages takes place every twelve years at the great *Kumbha Mela* festival in the city of Allahabad on the banks of the River Ganges to bathe in the river. Hindus believe that the Ganges River water washes away their sins. This is the largest gathering of people on the planet held during *Kumbha Mela*.

Judaism
The history of Judaism goes back to four thousand years. This is the oldest monolithic religion based on the belief of one god. Jews trace their history to the group of people called Hebrews, who lived in the region of what is now the Middle East. Historically, they were nomadic people, and they did not have a permanent home but moved from place to place. Jews believe that God appointed the Jews to be his chosen people to set an example of holiness and ethical behavior to the world. According to the Tanakh (Hebrew Bible), God promised Abraham, father of the Jewish people, to make of his offspring a great nation. Many generations later, he commanded the nation of Israel to love and worship only one god.

It is written in the Jewish scriptures how Abraham and his elderly wife, Sarah, longed for children. One night, Abraham heard God's voice that he would have as many descendents as there were stars in the sky and that they would live in a land of their own—the Promised Land. Eventually, Abraham's son, Isaac, was born. Later, Isaac had a son named Jacob, who was Israel named by God, and the descendants of Abraham were known as the Israelites.

Nearly 1250 BCE, the Israelites were freed from the pharaoh of Egypt, where they were slaves. According to Jewish scriptures, God chose Moses to plead with the Egyptian pharaoh king to free the Israelites. But when Pharaoh refused Moses's request, God sent several plagues. In the final plague, all the firstborn sons of the Egyptians and pharaohs were killed. The pharaoh king let the Israelites leave Egypt. When they started their journey, Pharaoh changed his mind and sent his army to chase in violent pursuit. When the Israelites reached the Red Sea, the waters miraculously divided in two parts, making a safe, dry path for Israelites. But when Pharaoh's army began to cross, the waters closed up, and all the horses and men drowned as a result. After the Exodus, when the Israelites were living in the desert, God sent the Ten Commandments. They are the following:

- I am the Lord, your god. Worship no god but me.
- Do not make or worship images.
- Do not use God's name for evil purposes.
- Observe the Sabbath (holy day).
- Respect your father and mother.
- Do not kill.
- Do not be unfaithful in marriage.
- Do not steal.
- Do not accuse anyone falsely.
- Do not envy other people's possessions.

The Hebrew Bible, the Tanakh, is a collection of twenty-four books. The first five books are known as the *Torah*, which means teachings. The instructions in the Torah were given by God to Moses. The remaining nineteen books are divided into the *Nevi'im*, which is composed of eight books, and the *Ketuvim*, which is composed of eleven books. All these nineteen books contain histories, prophecies, hymns, poems, and sayings. These books were written over about nine hundred years from around 1000 to 100 BCE, and they were written mainly in the Hebrew language. Today, the Jewish teaching is often referred to as the Torah. Another set of writings, known as *Talmud*, contains the thoughts of about two thousand rabbis. The Talmud has two parts—the *Mishnah* and the *Gemara*. The *Mishnah* is a collection of Jewish laws, and the *Gemara* explains the meanings and comments about the *Mishnah*. The *Midrash* is another important collection of the texts, which are stories that explain various aspects of *Tenaka*.

Jews are divided into two main groups—Orthodox and non-Orthodox. Orthodox Jews strap small boxes containing prayers around their head and around the arm nearest their heart before morning worship. Many Orthodox Jew men wear a hat or a skullcap called a *kippah* at that time. Orthodox Jews believe in Torah, which contains laws given by God to Moses. Non-Orthodox Jews think that human beings played a significant role in making the laws and bringing Judaism to modern life. Therefore, non-Orthodox Jews are considered Progressive and Liberal Jews.

Jewish worship is less restrictive than other religions. For example, people can join or leave the worship at any time and can talk quietly to one another during the worship service. The worship service contains readings from the Torah, hymns, and prayers. The worship is led by a rabbi, who also gives talk about Tanakh. In Orthodox synagogues, men and women are separated during worship. In Progressive synagogues, however, men and women sit together. Prayer is an important part of the Jewish faith, which can be said directly to God, especially to praise him. There are different formal written prayers to be said at different times, such as three times prayers each day, prayers for the Sabbath, and prayers for various festivals and fasts.

Buddhism

Buddhism was developed about 2,500 years ago by a man named Siddhartha Gautama. He was known as the Buddha, meaning the "Enlightened One." The Buddha was born a prince of the Shakya province in ancient India on the full moon day of May 563 BC. He lived over five hundred years before the birth of Jesus Christ. He gave up claim of the throne and left the kingdom in search of truth. Initially, he tried all the ascetic, extremist practices and reached a high level of tranquility and trance but did not find ultimate liberation. Eventually, after six years of struggle in the forest, including extreme fasting to the verge of near death, he discovered that the middle path to lead life (that is, neither indulging in luxury nor causing needless hardship to the body) was the best path. He attained enlightenment and wisdom, at the age of thirty-five, to find the answers of birth, death, suffering, and the end of sufferings. He taught to all for forty-five years until his death at the age of eighty in the year 483 BC. History tells that one evening, he sat down in the shade of a banyan tree near a temple of the Hindu god Vishnu at Bodh Gaya in India. He stayed there all night, deep in meditation. As dawn broke, he found the meaning of all things and became enlightened. When he achieved enlightenment, the Buddha attained *nirvana*, meaning freedom from the cycle of rebirth and freedom from suffering. Many scholars think that Buddhism is the starting point for a study of the evolution of modern scientific humanism.

The Buddha's main teachings are known as the Three Universal Truths, the Four Noble Truths, and the Eightfold Paths; together, all these are known as *dharma*. The following are the Universal Truths:

- Everything in life is impermanent and is constantly changing.
- Impermanence leads to suffering, which, in turn, makes life unsatisfactory. People become attached to things that cannot last, especially in the state of contentment. Knowing that the contentment must end is a source of suffering. Moreover, the suffering means not only the pain, anguish, and tragedies but also all those things that make life less than perfect.
- There is no unchanging personal self. Rather, self is a collection of changing characteristics. Self is like a chariot that is a collection of parts that are put together in a certain way and can be taken apart again.

The Four Noble Truths are the following:

- Life involves suffering.
- The cause of suffering is desire and attachment; in other words, suffering is caused by unreasonable expectations.
- Desire and attachment can be overcome; in other words, suffering ceases with the ceasing of unreasonable expectations.
- The way to overcome them or the way to reasonable expectations is to follow the Eightfold Middle Path.

The Eightfold Middle Paths are the following:

- right understanding of the Noble Truths, which are faith (in the teachings that are worth pursuing), knowledge (of the teachings and their compatibility to logic and science), experience (of the teachings through meditation), and wisdom (through enlightenment experience);

- right thought or intention (for example, trying to act considerately);
- right speech (for example, avoiding anger, lies, and gossip);
- right action (for example, living honestly and not harming living things);
- right work or livelihood (for example, abstaining from making a living that harms others);
- right effort (for example, trying hard to overcome desire and attachment);
- right mindfulness (for example, thinking before speech and action), in other words, to be diligently aware, mindful, and attentive with regard to activities of the body, sensations, or body feelings, activities of the mind, ideas, thoughts, conceptions, and emotions; and
- right concentration or meditation (for example, freeing the mind of distractions, leading to enlightenment and *nirvana*).

Buddhists take refuge and are united in their belief in the Triple Gem of *Buddha*, *dharma*, and *sangha*. The first part of the Triple Gem is *Buddha*. This refers to Buddha inside all of us, which includes Buddha-nature, our capacity for enlightenment, and the basic goodness in all people. Taking refuge in Buddha is not worship (especially in the typical use of god worship) to Buddha. At many Buddhist temples, people can be seen bowing and making prostrations to Buddha statues. This is just our respect, but not the same kind of worship that is done in other religions to a higher being. In many Asian countries, people greet one another with bows. They are not worshipping one another but just showing respect in the same way people shake hands to greet. The *dharma* refers to teachings of the Buddha. This gives comfort and solace as it is learned the *dharma* and its wonderful message. Sangha refers to the community of monks and nuns and gradually became known as the entire Buddhist community of monastic and laypeople.

The Buddha taught that the people have to take responsibility for their own actions for enlightenment. He laid out five rules or precepts that every Buddhist should obey in everyday life. They are to avoid harming living things,

- to avoid taking things that have not been freely given,
- to live a decent lifestyle,
- to avoid speaking unkindly or lying, and
- to avoid alcohol and drugs.

In addition to obeying the five precepts, it is important to do meditation, which is essential in achieving nirvana. Through meditation, one can search within the self, and a person can understand the truth of the Buddha's teachings.

Buddhism spread throughout Northern India. Emperor Ashoka helped Buddhism spread across India. His son Mahendra took the religion to Southern India and Sri Lanka, and it was spread east, along the trade routes into China. Buddhism was practiced in a befitting manner in India until 1200 CE, when Muslim invaders destroyed many Buddhist temples, shrines, and monasteries. After the death of Buddha, difference of opinion began, and it was divided into two branches of Buddhism as Theravada and Mahayana as a result. *Theravada* means "way or teachings of the elders." This group followed the original teachings of Buddha more closely than others. On the other hand, Mahayana Buddhists do not concentrate only on the Buddha Siddhartha Gautama. They also believe in *bodhisattvas*, which means "Buddha-to-be." Mahayana Buddhists use the *Tipitaka* and other more recent texts known as *sutras* as well. Mahayana Buddhists adapted to the new cultures they met, which led to the formation of distinctive branches such as Vajrayana, Pure Land, and Zen. Vajrayana Buddhism is also known as Tibetan Buddhism as it flourished in Tibet. Its leaders had political power as well as religious influence. Pure Land Buddhism began in China and spread farther to Japan in about the thirteenth century. The Pure Land Buddha is known as Amida Buddha. Zen Buddha also spread in China and Japan about the same time when Pure Land Buddhism spread. *Zen* means "meditation." Zen Buddhists use painting, martial arts, and specially designed gardens to help focus minds during meditation.

Christianity
Christianity was developed from Judaism in the first century CE. It is founded on the life, teachings, death, and resurrection of Jesus Christ, and those who follow him are called Christians. *Christ* is the Greek word for *Messiah*, meaning "anointed one." Jesus was born a Jew in Roman-occupied Judea (a place that is a part of Israel) about two thousand years ago. According to scriptures, God sent angel Gabriel to visit a young woman named Mary in Nazareth. Gabriel told Mary that she would be the mother of Jesus. Christians believe that Mary was a virgin despite being a mother.

Christianity has many different branches and forms with accompanying variety in beliefs and practices. The three major branches of Christianity are Roman Catholicism, Eastern Orthodoxy, and Protestantism, with numerous subcategories within each of these branches. Until the late twentieth century, most adherents of Christianity were in the West, though it has spread to every continent and is now the largest religion in the world. Traditional Christian beliefs include the belief in the one and only true god, who is one being and exists as Father, Son, and Holy Spirit, and the belief that Jesus is the divine and human Messiah sent to save the world. Christianity is also noted for its emphasis on faith in Christ as the primary component of religion.

The most important sermon that Jesus gave is known as the Sermon on the Mount. Jesus outlined his main teachings, and he blessed groups of people, including the meek, the peacemakers, and people who were being persecuted. He described God's law and said that people should obey them, for example, by loving their enemies and not judging others. Jesus said that people should repent for their sins (that is, actions against God's law) and make a fresh start. He also taught that love and serving others were more important than all the specifics of the Jewish law, and he spoke of God as Father. Jesus was charged with blasphemy, and he was handed over to the authorities by Judas, one of the disciples. Jesus was taken to the Roman governor named Pontius Pilate, who ordered for Jesus to be crucified. Two thieves were also crucified with him, one on each side. Jesus was buried in a tomb, but on the third day, after his crucifixion, the tomb was found empty. Christians believe that Jesus was resurrected and ascended to heaven to be united with God, his father.

The sacred text of Christianity is the Bible, divided in two parts—Old Testament, which is almost the same as the Jewish Tenaka, and the New Testament, the life of Jesus and how God could make a new relationship with all people through Jesus. It contains the story of the creation of the world, the history of the Jews, and the relationship between the Jews and God. The *New Testament* was written in Greek. It tells the life of Jesus and how God could make a new relationship with all people through Jesus. It also teaches that salvation comes through belief in the death and resurrection of Jesus. The New Testament is composed of twenty-seven books: four gospels, twenty-one epistles, and two books called Acts and Revelation.

Gospels were written by four evangelists: Matthew, Mark, Luke, and John. Each gospel describes the teaching, death, and resurrection of Jesus from the author's point of view. An epistle is a letter, and most of the epistles were written after the death of Jesus by a converted Jew, Paul. *Evangelist* means "announcer of good news." Each gospel describes the teaching, death, and resurrection of Jesus from the author's point of view. An *epistle* is a letter. Most of the epistles in the New Testament were written about thirty years after the death of Jesus by a converted Jew named Paul. He traveled throughout the Roman Empire, telling non-Jews about Jesus and building Christian communities known as churches. It is important to note that Paul's letters gave advice and encouragement to the early Christians. Acts (epistles) follows the gospels—the story after the resurrection of Jesus. Revelation, the final book, describes a vision of the end of time. Central to Christian practice is the gathering at churches for worship, fellowship, study, and engagement with the world through evangelism and social action.

After Jesus died, his teachings were spread by his followers. The first Christian sermon was given by Peter, one of the disciples. About three thousand people joined Christianity. The Christianity was further spread to Asia Minor, now modern Turkey, Greece, and to Rome by another disciple named Paul. In 313 CE, Roman emperor Constantine converted to Christianity, founded Constantinople near the old Greek city of Byzantine, and established Christianity.

By the sixteenth century, as many Christians were unhappy with the Roman Catholic Church, a new church known as Protestant was formed and pioneered by Martin Luther and John Calvin. They were the first who openly criticized the corrupt practices of the Catholic Church, which was very rich and powerful. This period was known as the Reformation period for Christianity. Catholics and Protestants hated each other, which led to intolerance, persecution, and religious wars.

In 1529, King Henry VIII of England ruled out the supreme authority of the pope and declared himself as the head of the church in England. Under Henry's son Edward VI, England was converted into a Protestant country, and many Catholics were murdered. Eventually, Henry's daughter Mary I brought back England into Catholicism, and many Protestants were killed during her reign. In 1558, Elizabeth, the queen of England, established the Church of England by making a compromise between Protestantism and Catholicism. Around the same time, many Protestants refused to conform to the established churches of northern Europe. During the seventeenth century, this group known as Nonconformist began to establish Nonconformist Churches. Nonconformists believed that worship should be very simple. Many of them do not consider themselves as Christians, and the mainstream Christians do not consider them as Christians. The Nonconformists are Baptists, Christian Scientists, Church of Christ, Congregationalists, Holiness Movement, Jehovah's Witnesses, Lutherans, Mennonites, Methodists, Mormons, Pentecostalists, Presbyterians, Quakers, Salvation Army, Seventh-Day Adventists, Unitarians, and United Reformed Church.

Christians had been following the instructions from Jesus to teach others about their faith since the first disciples. Thus, Christianity had been, and still is today, spread by ordinary Christians and by those who have devoted their lives to converting others to Christianity. The Christians who dedicate to spreading their belief to people in other places and countries are known as Christian missionaries. These missionaries were active in Africa and Asia, in particular, during the eighteenth and nineteenth centuries. Christian missionaries still work today in various parts of the world. Along with the teaching of Christian faith and building churches and missionaries, they help establish schools and hospitals in local areas where they work as missionaries.

Islam

The Arabic term *Islam* literally means "surrender" or "submission." Islam's believers (known as Muslims from the active participle of *Islam*) accept surrender to the will of Allah (the Arabic word for "God"). Allah is viewed as a unique god—creator, sustainer, and restorer of the world. The will of God, to which man is to submit, is made known through the Quran (or Koran), which was revealed to his messenger Muhammad. Followers of Islam are called Muslims, which mean "obedient ones." In Islam, Muhammad is considered the last of a series of prophets (including Adam, Noah, Abraham, Moses, Solomon, and Jesus), and his message simultaneously consummates and completes the "revelations" attributed to earlier prophets.

Muhammad was born in Makkah in Saudi Arabia in 570 CE. He was an orphan and was brought up by his uncle and became a camel driver and trader. He was unhappy about the lawlessness of his fellow people and was troubled by their belief in many gods. He used to go into the mountains to pray and think. At around the age of forty, while he was in a cave on Mount Nur, near Makkah, it is believed that God sent angel Gabriel to Muhammad to tell people of Makkah to worship only Allah (the Arabic word for "God") but not many gods. Later, Muhammad received messages from God through Gabriel, according to Islamic belief. He began to preach, and his followers grew in number, because of which the Makkahans feared about the popularity of Muhammad, and social and political unrest occurred as a result. In 622 CE, he and his followers migrated to Medina. In 629 CE, the Muslims were able to conquer Makkah, and Muhammad was finally accepted there as the prophet of Allah, and the community state of Islam emerged.

As mentioned earlier, Quran is the holy book for the Muslims. The Quran was compiled after several years of Muhammad's death from the revelations, which was memorized by his followers and passed on by word of mouth. The Quran teaches the Muslims that God is in control of everything that happens. Muslims try to follow and do the will of Allah in compliance to the words in the Quran rather than to follow an individual opinion through their life. Other than the Quran, the words and deeds of Muhammad (the Sunna) were collectively recorded in writings called Hadith (the teachings, deeds, and sayings of the Islamic prophet Muhammad), which helps interpret the Quran.

Islamic faith is founded on the five pillars—Shahadah, Salah, Zakah, Sawm, and Hajj.

Shahadah is the declaration of faith, which must be verbally said several times a day: "There is no god but Allah, and Muhammad is his messenger." Salah is the daily five-times prayers that need to be said in Arabic at dawn, just after midday, in the midafternoon, just after sunset, and after dark. These prayers consist of verses from the Quran praising Allah and asking for guidance. Zakah is the mandated duty for Muslims who can afford to give every year at least 2.5 percent of their savings and other valuables to the poor and needy. Sawm is a fasting ritual, which is to be followed during the ninth Islamic month, Ramadan. Muslims eat and drink nothing during the daylight hours. This is the month for studying the Quran and showing self-control and caring for others. Hajj is a pilgrimage to Makkah and visit Kaaba. This place of worship is thought to have been built by Ibrahim and one of his sons, Ismail. The Hajj is performed during the twelfth Islamic month. Special white garments are required to be worn during Hajj as a sign that all are equal during Hajj. It is mandatory for Muslims to perform Hajj at least once in a lifetime. However, the poor, old, and sick people are exempted from performing the Hajj.

There are two principal divisions of the Muslim population—Sunni and Shia. After Muhammad's death, his followers argued over the leadership in Islam. While one group argued that Abu Bakr would be the rightful successor of Muhammad, the other group said that Ali, Muhammad's cousin and son-in-law, would be the successor. The former group became Sunnis, and the latter group became Shias. Overwhelming majority of world's Muslims are Sunnis. Shia Muslim dominates in Iran and also exists in many other Muslim countries such as southern Iraq, Lebanon, and Bahrain. Both Sunnis and Shi'tes share the same views on Quran, Muhammad, and God, but they differ in some ways about the teachings of Islam into practice. Apart from these two groups in Islam, a relatively small number of Muslims (know as Sufis) seek to find God and directly aimed at gaining inner knowledge from him. This philosophy of Islam, called Sufism, emerged from about 800 CE.

In Islam, the religious laws come from the Quran and Sunna and are called the Sharia. It contains guidelines on matters related to a person's actions and affairs of state as well. It is important to note that Muslims living in non-Muslim countries are not in a position to keep the Sharia laws and, thus, torn apart between their desire to strictly follow Islamic laws and the customs of the country where they live. One of the important elements in Islam is *jihad*. The term *jihad* is used for a struggle and resist for causes, both religious and secular. The *Dictionary of Modern Written Arabic* defines the term as "fight, battle, jihad, holy war (against the infidels, as a religious duty)." Nonetheless, it is usually used in the religious sense, and its beginnings are traced back to the Quran and words and actions of Muhammad. In the Quran and in later Muslim usage, jihad is commonly followed by the expression *Fi sabil illah*, "In the path of God." The context of the Quran is elucidated by *Hadith*. Of the 199 references to *jihad* in perhaps the most standard collection of *Hadith*, *Bukhari*, all assume that *jihad* means "warfare."

The religion Islam was spread very rapidly after the death of Muhammad through remarkable successes, both by converting unbelievers to Islam and by military conquests of the opponents of Islamic communities. Expansion of the Islamic state was the desirable development because Muhammad himself had successfully spread and established the Islamic faith through conversion and conquest of those who stood against him. After Muhammad's death in 632, Abu Bakr (Muhammad's friend), as the first *caliph*, continued the effort to abolish paganism among the Arab tribes and also to incorporate Arabia into a region controlled by the political power of Medina. United by their faith in God and a commitment to political consolidation, the merchant elite of Arabia succeeded in consolidating their power throughout the Arabian Peninsula. Later, the *caliphs* waged many wars and began to launch some exploratory offensives north toward Syria. In 661, the Islamic capital was moved from Makkah to Damascus, and in 750, it was moved to Baghdad, and it remained there for the next five hundred years. In the eighth century, the Muslims conquered much of Spain and Portugal. Muslims ruled there until the late fifteenth century, when the Christians of Spain and Portugal joined forces and overthrew them. In the sixteenth and seventeenth centuries, Islamic emperies were very powerful in the regions of Turkey, Iran, and India, such as the Ottoman Empire, Safavid Empire, and Mogul Empire. By the end of the fifteenth century, most of the Christian Byzantine, including Constantinople (which was renamed as Istanbul), was conquered by the Muslims.

Sikhism

Sikhism was founded in the Punjab region in India by a man known as Guru Nanak in the fifteenth century CE (five hundred years ago) and is a monotheistic religion. Sikhs think religion should be practiced by living in the world and coping with life's everyday problems. The people who follow Sikhism are called Sikhs, which means "disciples." Guru Nanak passed on his enlightened leadership of this new religion to nine successive gurus. The final living guru, Guru Gobind Singh, died in 1708. Today, there are about twenty-three million Sikhs spread throughout the world; most of them live in India.

Guru Nanak was born in 1469 in the Talwandi village near Lahore in the Punjab. His parents were Hindus, but he grew up among Hindus and Muslims. Guru Nanak spent his last few years in the town of Kartarpur in Punjab. His followers or disciples formed a community that was friendly to everyone and opened a kitchen to feed the poor and needy. He died in 1539. Before his death, he chose one of his disciples named Lehna (the changed name was Angad) to succeed him. Angad developed a written script called *Gurmukhi* (meaning "from the mouth of the Guru"). Later, Sikh beliefs were passed down through a chain of gurus. During the Mogul rules in India, Sikhs suffered persecution by Muslims. Both the fifth and ninth gurus were killed for their beliefs. To protect and resist oppression and defend their faith, the tenth guru, Gobind Singh, founded the community of the *Khalsa* ("the Pure"). The last guru, Gobind Singh, decided not to select a guru as successor. Instead, he advised that the Sikh scriptures will guide all future followers of the Sikh religion. The scriptures were collected and composed in a book called the *Adi Granth*. Later, it was named as *Guru Granth Sahib* to show the book a high status. While the scriptures contain mainly of hymns written by the gurus, they also contain the writings of people from other faiths, including Muslims and Hindus.

Sikhs usually go to the temple to worship. The Sikhs temple is called *gurdwara*, meaning "door to the guru." However, Sikhs can also worship at home if they have a copy of the *Guru Granth Sahib*. Big *gurdwaras* are usually open all day and night to people of all religion. The *gurdwaras* provide meal in a dining room called a *langar* and have a place to sleep for anyone who needs it. Sikhs wear five symbolic items of dress; each begins with the letter *K*, known as five Ks. They are Kesh, Kangha, Kachera, Kara, and Kirpan. Although turbans worn by men are associated with the Sikh faith, it is not one of the five Ks. It is very much an admirable thing that, unlike many other religions, there are no divisions, branches, or groups in Sikhism; rather, it is a solid, unified religion with one identity.

Taoism and Confucianism

These two religions were developed in China in the sixth century BCE. Although religion is not important in China, Taoism and Confucianism have about ten million followers. *Taoism* is based on teachings written in the book called *Tao Te Ching*. The Tao stands for "the way." The followers of Taoism believe that the underlying spiritual force of the universe is present in all things but greater than all things. Taoists also believe that two opposite forces in the universe come from the *Tao—yin* and *yang*. The *yin* represents darkness and femininity, and *yang* represents brightness and masculinity. These two opposite qualities are believed to be the basis of all creation. The most important concept in Taoism is wu wei, which means acting naturally and not interfering with the process of life. Thus, a person should allow things to happen in the natural way rather than interfere or control events.

Confucianism was established by the Chinese philosopher named Confucius, who lived in the sixth century BCE. His real name was Kung Fu-Tzu, but later, he was known as Confucius, especially in the West. His teachings were intended for the rules of China. Later, his followers incorporated ideas from Taoism and Buddhism into Confucianism, which spread from China to Korea, Japan, Vietnam, and other Southeast Asian countries. Confucianism strongly emphasizes the importance of people's behavior. Realistically, Confucianism has not been considered as a religion because it places more importance on becoming a good citizen than on spirituality. In Confucianism, there is no priest, and followers worship Confucius as a great teacher but not a god. However, rituals are important to Confucians, which are performed and aimed at strengthening the five relationships (such as between rulers and subjects, father and son, elder brother and younger brother, husband and wife, and the relationship between two friends). One of the most important rituals is the worship of ancestors. According to Confucianism, soul exists after death, so it is important to worship the ancestors, especially on the anniversary of an ancestor's death.

Shinto

Shinto is an all-pervading, indefinable way that is quite universal. Shinto implies spontaneous following of the "way of the gods." Shinto is not really an ism. It is only a teaching. It is not a set of verbal theories or concepts. It is the all-pervading way.

Shinto religion was developed in Japan, which has over one hundred million followers. The main belief of Shinto religion is founded on the spiritual power called *kami* that exists in the natural world. While various types of *kami* are in Shinto, the most important one is the sun goddess called *Amaterasu*. This goddess is considered as the ruler of all spirits and the guardian of the people and the symbol of Japanese unity. It is also believed that all of Japan's emperors are the descendants of *Amaterasu*.

Nature is very sacred to Shinto followers. Thus, worships are performed in the beautiful natural settings called shrines. Followers visit shrines to pray for good fortune and to avoid evil spirits. They also believe that purity of body and spirit is important and that the key element of Shinto worship is purification, known as *harae*. Although there are no official sacred Shinto scriptures, several books are highly regarded by the followers. The main books are *Kojiki* and the *Nihon-gi*, which were written in the eighth century. Shinto shrines celebrate several festivals each year. The main festivals are *Haru Matsuri* (a spring festival), *Aki Matsuri* (a harvest festival), and *Rei-sai* (the major annual festival).

Purity is one of the fundamental virtues of Shinto ethics. There are two significations of purity. One is outer purity or bodily purity, and the other, inner purity or purity of heart. If a man is endowed with true inner purity of heart, he will surely attain god realization or communion with the divine. Sincerity is also the guiding ethical principle of Shinto. The system of Shinto resembles the system of Hinduism more than that of Confucianism or Buddhism. It is a kind of personal religion. It ascribes divine attributes to every being. It is a kind of pantheism. For the Japanese, nation means a harmonious complex of individuals, Kuni-hito. Salvation, for the Japanese, means the salvation of the whole nation instead of salvation of a few individuals.

The ten precepts of Shinto are the following:

- Do not transgress the will of the gods.
- Do not forget your obligations to ancestors.
- Do not offend by violating the decrees of the state.
- Do not forget the profound goodness of the gods through which calamity and misfortunes are averted and sickness is healed.
- Do not forget that the world is one great family.
- Do not forget the limitations of your own person.
- Do not become angry even though others become angry.
- Do not be sluggish in your work.
- Do not bring blame to the teaching.
- Do not be carried away by foreign teachings.

The religions above are the main ones among many others. Because of the limited scope of this book, I presented only a brief highlight of each of them. It is, however, recommended that readers thoroughly study to know their own religion. It's also a good idea to have a good knowledge about other religions as well.

Annexure 2

Conflict and Wars in the Recent Past

It is important to realize that most of the world's current "hot spots" have a complex interaction of economic, racial, ethnic, religious, and other factors. Below is the list of some conflicts that have as their base at least some degree of religious intolerance.

Country	Main Religious Groups Involved	Type of Conflict
Afghanistan	Extreme, radical fundamentalist Muslim terrorist groups and non-Muslims	Osama bin Laden headed a terrorist group called Al-Qaeda, headquartered in Afghanistan. They were protected by and integrated with the Taliban dictatorship in the country. The Northern Alliance of rebel Afghans, Britain, and the United States attacked the Taliban and Al-Qaeda, establishing a new regime in part of the country. The fighting continues.
Bosnia	Serbian Orthodox Christians, Roman Catholics, Muslims	Peace is holding because of the presence of peacekeepers.

Côte d'Ivoire	Muslims, Indigenous, Christians	After the elections in late 2000, government security forces "began targeting civilians solely and explicitly on the basis of their religion, ethnic group, or national origin. The majority of victims come from the largely Muslim north of the country or are immigrants or the descendants of immigrants." A military uprising continued the slaughter in 2002. It split the nation into two segments. Periods of peace and violence have alternated as the country struggles toward stability.
Cyprus	Christians and Muslims	The island is partitioned, creating enclaves for ethnic Greeks (Christians) and Turks (Muslims). A UN peacekeeping force is maintaining stability.
East Timor	Christians and Muslims	A Roman Catholic country. About 30 percent of the population died by murder, starvation, or disease after they were forcibly annexed by Indonesia (mainly Muslim). After voting for independence, many Christians were exter minated or exiled by the Indonesian army and army-funded militias in a carefully planned program of genocide and religious cleansing. East Timor won its independence from Indonesia in 2002. The situation there is now stable.
India	Animists, Christians, Hindus, Muslims, and Sikhs	Various conflicts heat up periodically, producing loss of life. Christians are regularly attacked in Orissa Province by militant Hindu extremists. Hindu-Muslim riots were very common, and still some scattered riots take place, although not significant.

Indonesia, Maluku Islands	Christians and Muslims	After centuries of relative peace, conflicts between Christians and Muslims started during 1999. About six thousand were killed. Over a half million people were internally displaced. Thousands were forced to convert to another religion. Peace talks were initiated by the government in early 2002. The situation appears to be stable.
Iraq	Kurds, Shiite Muslims, Sunni Muslims, Yazidi	This is a country with three main ethnic and religious groups: Shiites, Sunnis, and Kurds. For decades, one group has controlled the government, and the other two groups have suffered. In 2014, a new group invaded the country—ISIS. Its goal is to create a caliphate in the region, and it is now controlling large areas of Iraq and Syria.
Kashmir	Hindus and Muslims	A chronically unstable region of the world claimed by both Pakistan and India. The availability of stockpiles of nuclear weapons is destabilizing the region further. Thirty to sixty thousand people have died since 1989. A plebiscite would be the obvious solution, except that either Pakistan or India—whichever country polls show would lose the plebiscite—refuses to allow one.
Kosovo	Serbian Orthodox Christians and Muslims	Peace enforced by NATO peacekeepers. There is convincing evidence of past mass murder by Yugoslavian government (mainly Serbian Orthodox Christians) against ethnic Albanians (mostly Muslim).

Kurdistan	Primarily Alevis, Muslims, with Christians, Jews, Yarsan, and Yazidis	This is a country with a tenuous existence in parts of Iran, Iraq, Syria, and Turkey. It is involved heavily in a war with ISIS. Parts of northeast Syria are under Kurdish control and set up a government there. Some Kurds have autonomy in those areas where they form a majority. Others are attempting to establish Kurdistan as a separate company.
Macedonia	Macedonian Orthodox Christians and Muslims	Muslims (often referred to as ethnic Albanians) engaged in a civil war with the rest of the country, who are primarily Macedonian Orthodox Christians, during the 1990s. A peace treaty has been signed. Disarmament by NATO is complete.
Gaza	Jews, Muslims, and Christians	Israel and Gaza have experienced repeated cycles of tenuous peace followed by full-scale warfare. This has resulted in the deaths of thousands, mainly in Gaza. Major strife broke out in 2000 and again and again in 2014. Flare-ups continue. No resolution yet appears to be possible. Innocent people, including women and children in particular, suffer most.
Myanmar (formerly Burma)	Buddhists and Muslims	The Burmese public had experienced massive human rights abuses under a military dictatorship that was replaced in 2011 with a nominally civilian government. The Muslim Rohingya, a 4 percent to 10 percent minority (estimates differ), are suffering major oppression by the 80 percent Buddhist majority. Even though some of the Rohingya have lived in the country for three or more generations, the government withholds citizenship from them. About four hundred

		thousand (half of the total population of Rohingyas) have been expelled from Burma, who takes shelter in neighboring country Bangladesh.
Nigeria	Christians, Animists, and Muslims	The country is divided between Yoruba and Christians in the south who live in an uneasy peace with Muslims in the north. Meanwhile, there is a terrorist group called Boko Haram (roughly translated as "Western education is forbidden"). The country is struggling toward democracy after decades of Muslim military dictatorships. Meanwhile, Boko Haram is attempting to establish an Islamic state in the northern states.
Northern Ireland	Protestants and Catholics	After 3,600 killings and assassinations for over thirty years, a cease-fire is holding.
Pakistan	Sunni and Shiite Muslims	Low-level mutual attacks overshadowed by Taliban insurrectionists.
Philippines	Christians and Muslims	A low-level conflict between the mainly Christian central government and Muslims in the south of the country has continued for centuries.
Somalia	Wahhabi and Sufi Muslims	Sufi Muslims, a tolerant, moderate tradition of Islam, are fighting the Shabaab, who follow the Wahhabi tradition of Islam in a continuing conflict.
South Africa	Animists and "Witches"	Hundreds of persons suspected and accused as witches practicing black magic are murdered each year.
		Tamils (a mainly Hindu 18 percent minority) are involved in a war aimed at dividing the island and creating a homeland for themselves. Conflict had been underway since 1983 with

Sri Lanka	Buddhists and Hindus	the Sinhalese Buddhist majority (70 percent). Over a hundred thousand people have been killed. The conflict took a sudden change for the better in September 2002, when the Tamils dropped their demand for complete independence. The South Asian tsunami in December 2004 induced some cooperation. By 2009, the Tamil uprising was crushed by the government.
Sudan	Animists, Christians, and Muslims	There is a complex ethnic, racial, religious conflict in which the Muslim regime committed genocide against both animists and Christians in the south of the country. Slavery and near slavery were practiced. A cease-fire was signed in May 2006 among some of the combatants. Warfare continues in the Darfur region, primarily between a Muslim militia and Muslim inhabitants.
Thailand	Buddhists and Muslims	Muslim rebels have been involved in a bloody insurgency in Southern Thailand—a country that is 95 percent Buddhist.
Tibet	Buddhists and Communists	The country was annexed by Chinese Communists in the late 1950s. Brutal suppression of Buddhism continues.
Uganda	Animists, Christians, and Muslims	Christian rebels of the Lord's Resistance Army are conducting a civil war in the north of Uganda. Their goal is a Christian theocracy whose laws are based on the Ten Commandments. They abduct, enslave, and rape about two thousand children a year.

(Source: http://www.religioustolerance.org/curr_war.htm)

Annexure 3

Secular Humanist Principles[*]

1) There is no god and no deity; there are only us, the material world, and the ecosystem surrounding us. There is no soul and supernature; everything, including ourselves, is made of materials.

2) Things happen, not because of God's wish and design but because of the underlined natural principles and the randomness embedded in those principles.

3) There is an end to everything, from the cosmo to our individual lives. The values of things are not in their everlasting eternality. The values of things reside in their duration, in the process.

4) Our value systems and appreciation of things should shift from the infinity and eternality, which do not exist, to the transient moments that happen all around us.

5) There are three cornerstones for our humanist principles: science, which tells us how things work; evolution, which tells us where we came from and why we are the way we are; and the happiness principle, which tells us how we should conduct our life.

6) The purpose of life is to pursue happiness during our lives. A good life is a happy and exciting life. The ultimate measure of life's success is the happiness in our lives.

7) Everyone should have the right and mean to be happy in his/her life, regardless of his/her social status. Feeling happy is the most basic and fundamental human right.

---

[*] Source: http://purpose-of-life-1.blogspot.ca/2008/12/secular-humanist-principle.html

8) The way to feel happy is to satisfy our needs and desires, to satisfy our human nature. To have things we want to do and to do the things we want to do—that is how we should pursue our happiness.

9) While we should have a plan for the future, what is more important is to grasp the present and to live in the moment. We might not be able to grasp the future because of the randomness of things, but we can always grasp the present.

10) Things do not have meanings, good or bad; we give them the meaning. If something makes us sad and we cannot change the thing itself, we should change our own minds. We should follow the *yuan* (the natural way of things) and accept our own fate (the random things happening to us). Changing our desire, taking one step back, things will not be as bad as we thought.

11) Our societal value system and our ethical and moral standard should be built to serve human beings, to enhance the human happiness. Our social institutions and the way of life, including the marriage system, should be constructed to maximize our personal happiness.

12) Human species is part of the earth's ecosystem. It is our responsibility and also for our own benefit to maintain this ecosystem.

13) Modern human civilization is built on social networking, organization, and labor specialization. This civilization is the basis of many of our happiness. To maintain this civilization and social order, it might be necessary to suppress some of our evolutionarily formed human nature like violence, hatred, and jealousy. This suppression is a sacrifice based on our value choice; it is not a suppression of absolute evil.

14) There is no sin, no evil, and no absolute good or bad. The merit of one thing and its moral values should be measured by its service to human happiness while maintaining the earth ecosystem and the necessary social order.

15) The most valuable things for human beings are truth, beauty, and love. We should promote human love while suppressing hatred and jealousy. It is love that brings us the most happiness.

16) Everyone has the right to conduct his/her own way of life as long as it does not constitute a direct physical harm to other people. We should accept and tolerate different ways of life and view diversity as a merit of a society.

17) We individuals are always a part of a larger existence. This larger existence—should it be family, community, nation, human species, earth ecosystem, or personal projects and societal endeavors—gives

our lives a more enduring meaning and provides continuity beyond our own existence.

18) For us individuals, there is nothing beyond our own death. We just do not exist there; our existences have a finite range in both space and time. It doesn't make logical sense to talk about what happens to us after we die. It only makes sense to talk about what happens to us while we are still alive.

19) The society and the larger existence, for which we are part of, will still exist beyond our individual death. As we care about ourselves, we also care about this larger existence, because we are part of it and it is part of us. Thus, we care about the time after our own death, and this care provides a continuity of life.

20) For us individuals, our finite existence is a mixed blessing. On the one hand, death makes us sad because of our unfinished endeavors. On the other hand, death is a relief. It is tranquilizing to think no matter how good or bad things are right now, there is always an end. It is also comforting to realize things, experiences, and feelings will be continued by other individuals. We should celebrate death as we celebrate the completion of a project—the project of our life. A life is good as long as it is happy, regardless of its length.

21) To do the best we can do, to accept the results as our fate, to enjoy the moment as we can grasp, and to face death with calm—that is how we should conduct our life as a secular humanist.

Annexure 4

Vivid Horrific Cases of Honor Killings[*]

- A man who rapes his own daughter and then, when she becomes pregnant, kills her to save the "honor" of his family. Or the Turkish father and grandfather of a sixteen-year-old girl, Medine Mehmi, in the province of Adiyaman, who was buried alive beneath a chicken coop in February for "befriending boys." Her body was found forty days later in a sitting position and with her hands tied.

- Aisha Ibrahim Duhulow, thirteen, who, in Somalia, in 2008, in front of a thousand people, was dragged to a hole in the ground—all the while screaming, "I'm not going! Don't kill me!"—then buried up to her neck and stoned by fifty men for adultery. After ten minutes, she was dug up, found to be still alive, and put back in the hole for further stoning. What was her crime? She had been raped by three men, and, fatally, her family decided to report the facts to the Al-Shabaab militia that ran Kismayo. Or the Islamic Al-Shabaab "judge" in the same country who announced the 2009 stoning to death of a woman—the second of its kind the same year—for having an affair.

- A young woman was found in a drainage ditch near Daharki in Pakistan, honor killed by her family as she gave birth to her second

[*] Source: http://www.independent.co.uk/voices/commentators/fisk/robert-fisk-the-crimewave-that-shames-the-world-2072201.html.

child—her nose, ears, and lips chopped off before being axed to death, her first infant lying dead among her clothes, her newborn's torso still in her womb, its head already emerging from her body. She was badly decomposed; the local police were asked to bury her. Women carried the three to a grave, but a Muslim cleric refused to say prayers for her because it was "irreligious" to participate in the namaz-e-janaza prayers for "a cursed woman and her illegitimate children."

- Munawar Gul shot and killed his twenty-year-old sister, Saanga, in the North-West Frontier Province of Pakistan, along with the man (Aslam Khan) he suspected was having "illicit relations" with her.

- In August 2008, five women were buried alive for "honor crimes" in Balochistan by armed tribesmen; three of them—Hameeda, Raheema, and Fauzia—were teenagers who, after being beaten and shot, were thrown still alive into a ditch where they were covered with stones and earth. When the two older women, aged forty-five and thirty-eight, protested, they suffered the same fate. The three younger women had tried to choose their own husbands. In the Pakistani parliament, the MP Israrullah Zehri referred to the murders as part of a "centuries-old tradition," which he would "continue to defend."

- In December 2003, a twenty-three-year-old woman in Multan, identified only as Afsheen, was murdered by her father because, after an unhappy arranged marriage, she ran off with a man called Hassan, who was from a rival, feuding tribe. Her family was educated—they included civil servants, engineers, and lawyers. "I gave her sleeping pills in a cup of tea and then strangled her with a dopatta [a long scarf, part of a woman's traditional dress]," her father confessed. He told the police, "Honor is the only thing a man has. I can still hear her screams. She was my favorite daughter. I want to destroy my hands and end my life." The family had found Afsheen with Hassan in Rawalpindi and promised she would not be harmed if she returned home. They were lying.

- Zakir Hussain Shah slit the throat of his daughter Sabiha, eighteen, at Bara Kau, in June 2002, because she had "dishonored" her family. But under Pakistan's notorious qisas law, heirs have powers to pardon a murderer. In this case, Sabiha's mother and brother "pardoned" the father, and he was freed. When a man killed his four sisters in Mardan in the same year because they wanted a share of his inheritance, his mother "pardoned" him under the same law. In Sargodha, around the same time, a man opened fire on female members of his family, killing two of his daughters. Yet again, his wife—and several other daughters wounded by him—"pardoned" the murderer because they were his heirs.

- Outrageously, rape is also used as a punishment for "honor" crimes. In Meerwala village in Punjab, in 2002, a tribal "jury" claimed that an eleven-year-old boy from the Gujar tribe, Abdul Shakoor, had been walking with a thirty-year-old woman from the Mastoi tribe, which "dishonored" the Mastois. The tribal elders decided that to "return" honor to the group, the boy's eighteen-year-old sister, Mukhtaran Bibi, should be gang-raped. Her father, who was warned that all the female members of his family would be raped if he did not bring Mukhtar to them, dutifully brought his daughter to this unholy "jury." Four men, including one of the "jury," immediately dragged the girl to a hut and raped her, while up to a hundred men laughed and cheered outside. She was then forced to walk naked through the village to her home. It took a week before the police even registered the crime as a "complaint."

- In 2001, a Karachi man, Bilal Khar, poured acid over the face of his wife, Fakhra Yunus, after she left him and returned to her mother's home in the red-light area of the city. The acid fused her lips, burned off her hair, melted her breasts and an ear, and turned her face into "a look of melted rubber."

- In the same year, a twenty-year-old woman called Hafiza was shot twice by her brother, Asadullah, in front of a dozen policemen outside a Quetta courthouse because she had refused to follow the tradition of marrying her dead husband's elder brother. She had then married another man, Fayaz Moon, but police arrested the girl and brought her back to her family in Quetta on the pretext that the couple could formally marry there. But she was forced to make a claim that Fayaz had kidnapped and raped her. It was when she went to court to announce that her statement was made under pressure—and that she still regarded Fayaz as her husband—that

Asadullah murdered her. He handed his pistol to a police constable who had witnessed the killing.

- In 1999, a mentally retarded sixteen-year-old girl, Lal Jamilla Mandokhel, was reportedly raped by a junior civil servant in Parachinar in the North-West Frontier Province of Pakistan. Her uncle filed a complaint with the police but handed Lal over to her tribe, whose elders decided that she should be killed to preserve tribal "honor." She was shot dead in front of them. In the same year, Arbab Khatoon was raped by three men in the Jacobabad District. She filed a complaint with the police. Seven hours later, she was murdered by relatives, who claimed she had "dishonored" them by reporting the crime.

- In 2006, authorities in the Kurdish area of Southeast Anatolia reported that a woman tried to commit suicide every few weeks on the orders of her family. Others were stoned to death, shot, buried alive, or strangled. A seventeen-year-old woman called Derya, who fell in love with a boy at her school, received a text message from her uncle on her mobile phone. It read, "You have blackened our name. Kill yourself and clean our shame, or we will kill you." Derya's aunt had been killed by her grandfather for an identical reason. Derya tried to carry out her family's wishes. She jumped into the Tigris River, tried to hang herself, and slashed her wrists—all to no avail. Then she ran away to a women's shelter.

- It took thirteen years before Murat Kara, forty, admitted in 2007 that he had fired seven bullets into his younger sister after his widowed mother and uncles told him to kill her for eloping with her boyfriend. Before he murdered his sister in the Kurdish city of Diyarbakir, neighbors had refused to talk to Murat Kara, and the imam said he was disobeying the word of God if he did not kill his sister. So he became a murderer. Honor restored.

- A Turkish journalist, Mehmet Farac, recorded the "honor" killing of five girls in the late 1990s in the province of Sanliurfa. Two of them—one was only twelve—had their throats slit in public squares, two others had tractors driven over them, and the fifth was shot dead by her younger brother. One of the women who had her throat cut was called Sevda Gok. Her brothers held her arms down as her adolescent cousin cut her throat.

- In 2001, Sait Kina stabbed his thirteen-year-old daughter to death for talking to boys in the street. He attacked her in the bathroom with an ax and a kitchen knife. When the police discovered her corpse, they found that the girl's head had been so mutilated that the family had tied it together with a scarf. Sait Kina told the police, "I have fulfilled my duty."

- In the same year, an Istanbul court reduced a sentence against three brothers from life imprisonment to between four and twelve years after they threw their sister to her death from a bridge after accusing her of being a prostitute. The court concluded that her behavior had "provoked" the murder. For centuries, virginity tests have been considered a normal part of rural tradition before a woman's marriage. In 1998, when five young women attempted suicide before these tests, the Turkish family affairs minister defended mandated medical examinations for girls in foster homes.

- British Kurdish Iraqi campaigner Aso Kamal of the Doaa Network against Violence believes that between 1991 and 2007, 12,500 women were murdered for reasons of "honor" in the three Kurdish provinces of Iraq alone—350 of them in the first seven months of 2007, for which there were only five convictions. Many women were ordered by their families to commit suicide by burning themselves with cooking oil. In Sulimaniya Hospital in 2007, surgeons were treating many women for critical burns that could never have been caused by cooking "accidents" as the women claimed.

- One patient, Sirwa Hassan, was dying of 86 percent burns. She was a Kurdish mother of three from a village near the Iranian border. In 2008, a medical officer in Sulimaniya told the AFP news agency that in May alone, fourteen young women had been murdered for "honor" crimes in ten days.

- In Tikrit, Iraq, a young woman in the local prison sent a letter to her brother in 2008, telling him that she had become pregnant after being raped by a prison guard. The brother was permitted to visit the prison, walked into the cell (where his pregnant sister was held), and shot her dead to spare his family "dishonor." The mortuary in Baghdad took DNA samples from the woman's fetus and also from guards at the Tikrit prison. The rapist was a police lieutenant colonel. The reason for the woman's imprisonment was unclear. One report said the colonel's family had "paid off" the woman's relatives to escape punishment.

- In Basra, in 2008, police reported that fifteen women a month were being murdered for breaching "Islamic dress codes." One seventeen-year-old girl, Rand Abdel-Qader, was beaten to death by her father two years ago because she had become infatuated with a British soldier. Shawbo Ali Rauf, nineteen, was taken by her family to a picnic in Dokan and shot seven times because they had found an unfamiliar number on her mobile phone.

- In Nineveh, Du'a Khalil Aswad was seventeen when she was stoned to death by a mob of two thousand men for falling in love with a man outside her tribe.

- In Jordan, women's organizations say that per capita, the Christian minority in this country of just over five million people are involved in more "honor" killings than Muslims—often because Christian women want to marry Muslim men. But the Christian community is loath to discuss its crimes, and the majority of known cases of murder are committed by Muslims. Their stories are wearily and sickeningly familiar. Sirhan, in 1999, boasted of the efficiency with which he killed his young sister, Suzanne. Three days after the sixteen-year-old had told police she had been raped, Sirhan shot her in the head four times. "She committed a mistake, even if it was against her will," he said. "Anyway, it's better to have one person die than to have the whole family die of shame."

- In 2001, a twenty-two-year-old Jordanian man strangled his seventeen-year-old married sister—the twelfth murder of its kind in seven months—because he suspected her of having an affair. Her husband lived in Saudi Arabia. In 2002, Souad Mahmoud strangled his own sister for the same reason. She had been forced to marry her lover—but when the family found out she had been pregnant before her wedding, they decided to execute her.

- In 2005, three Jordanians stabbed their twenty-two-year-old married sister to death for taking a lover. After witnessing the man enter her home, the brothers stormed into the house and killed her. They did not harm her lover.

- By March 2008, the Jordanian courts were still treating "honor" killings leniently. That month, the Jordanian criminal court sentenced two men for killing close female relatives "in a fit of fury" to a mere six months and three months in prison. In the first case, a husband had found a man in his home with his wife and suspected she was having an affair. In the second, a man shot dead his twenty-nine-year-old married sister for leaving home without her husband's consent and "talking to other men on her mobile phone." In 2009, a Jordanian man confessed to stabbing his pregnant sister to death because she had moved back to her family after an argument with her husband; the brother believed she was "seeing other men."

- Three men in Amman stabbed their forty-year-old divorced sister fifteen times last year for taking a lover; a Jordanian man was charged with stabbing to death his daughter, twenty-two, with a sword because she was pregnant outside wedlock. Many of the Jordanian families were originally Palestinian. Nine months ago, a Palestinian stabbed his married sister to death because of her "bad behavior." But last month, the Amman criminal court sentenced another sister killer to ten years in prison, rejecting his claim of an "honor" killing—but only because there were no witnesses to his claim that she had committed adultery.

- In Palestine itself, Human Rights Watch has long blamed the Palestinian police and justice system for the near-total failure to protect women in Gaza and the West Bank from "honor" killings. For example, a seventeen-year-old girl was strangled by her older brother in 2005 for becoming pregnant—by her own father. He was present during her murder. She had earlier reported her father to the police. They neither arrested nor interrogated him.

- In the same year, some masked Hamas gunmen shot dead a twenty-year-old, Yusra Azzami, for "immoral behavior" as she spent a day out with her fiancé. Azzami was a Hamas member; her husband-to-be, a member of Fatah. Hamas tried to apologize and called the dead woman a martyr—to the outrage of her family. Yet only last year, long after Hamas won the Palestinian elections and took over the Gaza Strip, a Gaza man was detained for bludgeoning his daughter to death with an iron chain because he discovered she owned a mobile phone on which he feared she was talking to a man outside the family. He was later released.

- In Lebanon (known as liberal country), there are occasional "honor" killings. The most notorious one was that of a thirty-one-year-old woman, Mona Kaham, whose father entered her bedroom and cut her throat after learning she had been made pregnant by her cousin. He walked to the police station in Roueiss in the southern suburbs of Beirut with the knife still in his hand. "My conscience is clear," he told the police. "I have killed to clean my honor." Unsurprisingly, a public opinion poll showed that 90.7 percent of the Lebanese public opposed "honor" crimes. Of the few who approved of them, several believed that it helped limit interreligious marriage.

- In Syria, Lubna, a seventeen-year-old living in Homs, was murdered by her family because she fled to her sister's house after refusing to marry a man they had chosen for her. They also believed—wrongly—that she was no longer a virgin.

- Tribal feuds often provoke "honor" killings in Iran and Afghanistan. In Iran, for example, a governor's official in the ethnic Arab province of Khuzestan stated in 2003 that forty-five young women under the age of twenty had been murdered in "honor" killings in just two months, none of which brought convictions. All were slaughtered because of the girl's refusal to agree to an arranged marriage, failing to abide by Islamic dress code, or being suspected of having contacts with men outside the family.

- Through the dark veil of Afghanistan's village punishments, we glimpse just occasionally the terror of teenage executions. When Siddiqa, who was only nineteen, and her twenty-five-year-old fiancé, Khayyam, were brought before a Taliban-approved religious court in Kunduz Province, their last words were "We love each other, no matter what happens." In the bazaar at Mulla Quli, a crowd—including members of both families—stoned Siddiqa to death first and then Khayyam.

- A woman identified as Bibi Sanubar, a pregnant widow, was lashed a hundred times and then shot in the head by a Taliban commander. In April of last year, Taliban gunmen executed by firing squad a man and a girl in Nimruz for eloping when the young woman was already engaged to someone else. History may never disclose how many hundreds of women—and men—have suffered a similar fate at the hands of deeply traditional village families or the Taliban.

• But the contagion of "honor" crimes has spread across the globe, including acid attacks on women in Bangladesh for refusing marriages. In one of the most terrible Hindu "honor" killings in India this year, an engaged couple, Yogesh Kumar and Asha Saini, were murdered by the nineteen-year-old bride-to-be's family because her fiancé was of lower caste. They were apparently tied up and electrocuted to death. A similar fate awaited eighteen-year-old Vishal Sharma, a Hindu Brahmin, who wanted to marry Sonu Singh, a seventeen-year-old Jat—an "inferior" caste that is usually Muslim. The couple was hanged, and their bodies burned in Uttar Pradesh.

• In Chechnya, Russia's chosen president, Ramzan Kadyrov, has been positively encouraging men to kill for "honor." When seven murdered women were found in Grozny, shot in the head and chest, Kadyrov announced—without any proof but with obvious approval—that they had been killed for living "an immoral life." Commenting on a report that a Chechen girl had called the police to complain of her abusive father, he suggested the man should be able to murder his daughter. "If he doesn't kill her, what kind of man is he? He brings shame on himself!"

• In the "West" as well, immigrant families have sometimes brought their baggage and the cruel traditions of their home. An Azeri immigrant was charged in Saint Petersburg for hiring hit men to kill his daughter because she "flouted national tradition" by wearing a miniskirt near the Belgian city of Charleroi. Sadia Sheikh was shot dead by her brother Moussafa because she refused to marry a Pakistani man chosen by her family. In the suburbs of Toronto, Kamikar Kaur Dhillon slashed his Punjabi daughter-in-law, Amandeep, across the throat because she wanted to leave her arranged marriage, perhaps for another man. He told Canadian police that her separation would "disgrace the family name."

- In Britain, Surjit Athwal, a Punjabi Sikh woman, was murdered by the orders of her London-based mother-in-law for trying to escape a violent marriage. A fifteen-year-old girl, Tulay Goren, a Turkish Kurd from north London, was tortured and murdered by her Shia Muslim father because she wished to marry a Sunni Muslim man. A girl named Heshu Yones, sixteen, was stabbed to death by her father in 2005 for going out with a Christian boy. In Accrington, Caneze Riaz, along with her four children (the youngest one was ten years old), was burned alive by her husband because of their "Western ways."

These are just a few horror stories to prove the pervasive, spreading disease of what is to be recognized as a crime to modern civilization. It's a tradition of family savagery that brooks no merciful intervention, no state law, rarely any remorse by the society where it belongs.

More Examples
Burning, cutting, mutilating, raping, and abusing a woman are perfectly legal and fine in Islam. Burning or damaging the Koran is not.

Ghazala Khan (October 29, 1986–September 23, 2005) was a Pakistani woman who was shot and killed in Denmark in broad daylight by her brother after she had married against the will of the family. The murder of Ghazala had been ordered by her father to save the "family honor."[*]

Honor killings are common in Turkey. Spousal abuse and murder is the number one cause of death for women in Turkey, not cancer, diabetes, or heart disease.[**]

Middle East Quarterly: 91 percent of honor killings are committed by Muslims worldwide.[***]

Pew Research (2013): Large majorities of Muslims favor Sharia. Among those who do, stoning women for adultery is favored by 89 percent

[*] Source: https://muslimstatistics.wordpress.com/2014/10/04/muslim-honor-killings-statistics/

[**] Source: https://muslimstatistics.wordpress.com/2014/10/04/muslim-honor-killings-statistics/

[***] Source: http://www.canadafreepress.com/index.php/article/43207

in Pakistan, 85 percent in Afghanistan, 81 percent in Egypt, 67 percent in Jordan, 50 percent in "moderate" Indonesia, Malaysia, and Thailand, 58 percent in Iraq, 44 percent in Tunisia, 29 percent in Turkey, and 26 percent in Russia.*

The UNFP estimated that approximately as many as five thousand women are murdered in the name of honor killing each year worldwide. However, that is low estimate, as reports from Turkey, Jordan, Pakistan, and the Palestinian territories, among other locales, are filtering in at an alarming rate. Women's groups in the Middle East and Southwest Asia suspect the victims are at least four times the United Nations' latest world figure of around five thousand deaths a year. Nevertheless, statistics in Germany, Sweden, other parts of Europe, the United Kingdom, Canada, and the United States suggest that young Muslim women are becoming increasingly vulnerable.**

---

* Source: http://www.pewforum.org/uploadedFiles/Topics/Religious_Affiliation/Muslim/worlds-muslims-religion-politics-society-full-report.pdf

** Source: https://wikiislam.net/wiki/Muslim_Statistics_-_Honor_Violence

BIBLIOGRAPHY

Baggini, Julian. *Atheism: A Very Short Introduction*. Oxford: Oxford University Press, 2003.http://en.wikipedia.org/wiki/secular_morality.

Barbour, Ian. "Science and Religion, Models and Relations." In *Encyclopedia of Science and Religion*. Encyclopedia.com, 2003. Accessed January 19, 2016. http://www.encyclopedia.comhttp://www.encyclopedia.comhttp://www.encyclopedia.com.

Bees, Rev. H. May. 2015. https://consortiumnews.com/2015/05/03/letting-scientific-knowledge-into-religion/.Bostrom, Nick. "A History of Transhumanist Thought Faculty of Philosophy." *Oxford University Journal of Evolution and Technology* 14 (April 2005).http://jetpress.org/volume14/freitas.html

Boyer, Pascal. *Religion Explained*. London, UK: William Heinemann, Random House Group Ltd., 2001. www.humanreligions.info/secular_moral.html/.Burns, Kevin. *Eastern Philosophy*. Canada: Arcturus, 2006.

Cordeiro, José (www.cordeiro.org) studied engineering at MIT, Cambridge; economics at Georgetown University, Washington; management at INSEAD in France; and science at Universidad Simón Bolívar in Venezuela.

Davies, Rev. A. Powell. http://dmuuc.org/aboutworship/dr-a-powell-davies-bio-sermons/can-science-religion-together/.Dawkins, Richard. *The God Delusion*. New York: A Mariner Book, Houghton Mifflin Company, 2006.

Dennett, Daniel. *Breaking the Spell: Religion as a Natural Phenomenon.* New York: Viking, 2006.

———. "The Scientific Study of Religion." *Point of Inquiry* (2011). Discussion of morality starts. http://en.wikipedia.org/wiki/secular_morality.

Donnelly, Jack. *Universal Human Rights in Theory and Practice.* 3rd ed. Cornell University Press, 2013. Cited in www.humanreligions.info/secular_moral.html/.

Epstein, Greg M. *Good without God: What a Billion Nonreligious People Do Believe.* New York: Harper Collins, 2009. Cited in http://en.wikipedia.org/wiki/secular_morality.———. *Good without God: What a Billion Nonreligious People Do Believe.* New York: Harper Collins, 2009.Grudin, Robert. *The Most Amazing Thing.* 2001. Cited in http://en.wikipedia.org/wiki/Humanism.

Harris, Sam. *The End of Faith: Religion, Terror, and the Future of Reason.* London: Simon & Schuster, 2006. Cited in http://www.hts.org.za/index.php/HTS/article/view/. Hawking, Stephen, and Leonard Mlodinow. *The Grand Design.* New York: Bantam Books, an imprint of the Random House Publishing Group, 2010.

Hitchens, Christopher. *God Is Not Great—How Religion Poisons Everything.* Toronto, Ontario, Canada: McClelland & Stewart, 2007. Cited in *Why I Became an Atheist.* John W. Loftus. Amhert, New York: Prometheus Books, 2012.———. "Hitchens—The Morals of an Atheist; Unknown Knowledge." YouTube talk in August 2007. Retrieved in January 4, 2010. Cited in http://en.wikipedia.org/wiki/secular_morality.Hockey, David. *Developing a Universal Religion.* Portland, Ontario, Canada: Stephenson-Hockey Publishing, 2003. Source: https://upload.wikimedia.org/wikipedia/commons/5/56/Developing_a_Universal_Religion_Part.

Human Rights Watch. "A Question of Security: Violence against Palestinian Women and Girls." *Human Rights Watch* 18 (7):1–100, 2006. Hutcheon, Pat Duffy. *The Road to Reason.* Ottawa Canada: Canadian Humanist Publications, 2001.

Paul, Gregory S. "Cross-National Correlations of Quantifiable Societal Health with Popular Religiously and Secularism in the Prosperous

Democracies: A First Look." *Journal of Religion and Society*. Baltimore, Maryland: 2005. Cited in http://en.wikipedia.org/wiki/secular_morality.

"Interpreting Honor Crimes: The Institutional Disregard Towards Female Victims of Family Violence in the Middle East." *International Journal of Criminology and Sociological Theory* 3, no. 1 (June 2010). Kaku, Michio. *Physics of the Future*. New York: Doubleday, 2011.Lamont, Corliss. *The Philosophy of Humanism*. Amherst, NY: Humanist Press, a division of the American Humanist Association, 1997. http://www.corliss-lamont. org/philos8.pdf.

Marcus, Gary. *Kluge: The Haphazard Evolution of the Human Mind*. New York: Mariner Books, 2009. Cited in *Why I Became an Atheist*. John W. Loftus. Amhert, New York: Prometheus Books, 2012.

Mathew, R. M. "End of Islam, Hinduism and Christianity and Rise of the Age of Humanism, Spirituality and the Universal God—Sanadhana Dharma." 2015.Morgan, Michael L. *The Essential Spinoza: Ethics and Related Writings*. Indianapolis/Cambridge: Hackett Publishing Company Inc., 2006.

Murry, William R. Review of *The Psychological Roots of Religious Belief: Searching for Angels and the Parent-God*, by M. D. Faber. http://www. meadville.edu/uploads/files/152.pdf (2009).Nietzsche, Friedrich. *Human, All Too Human: A Book for Free Spirits*. New York: Cambridge University Press, 1986.Nuwer, Rachel. Posted in Future, BBC, December 19, 2014. Roden, David. *Posthuman Life: Philosophy at the Edge of the Human*. Routledge, 2015. http://philosophicaldisquisitions.blogspot.ca/2015/07/ humanism-transhumanism-and-speculative.html.

Rosenberg, Alex. *The Atheist's Guide to Reality: Enjoying Life without Illusion*. New York: W. W. Norton & Company Inc, 2011.

Shermer, Michael. *Why People Believe Weird Things*. 2nd ed. New York: Henry Holt, 2002. Cited in *Why I Became an Atheist*. John W. Loftus. Amhert, New York: Prometheus Books, 2012.

Stenger, Victor J. *God: The Failed Hypothesis—How Science Shows that God Does Not Exist*. Prometheus Books, 2007. Cited in www. humanreligions.info/secular_moral.html/.Stock, Gregory. *Redesigning*

Humans: Our Inevitable Genetic Future. New York, NY: Houghton Mifflin Company, 2002.

Tavris, Carol, and Elliot Aronson. *Mistakes Were Made (Bad Not by Me): Why We Justify Foolish Beliefs, Bad Decisions, and Hurtful Acts.* Orlando, Florida: Harcourt, 2007. Cited in *Why I Became an Atheist.* John W. Loftus. Amhert, New York: Prometheus Books, 2012.Ukuekpeyetan-Agbikimi, Nathaniel Aminorishe. *Global Journal of Arts Humanities and Social Sciences* 2, no.7 (2013).

Winston E. Langley, 2007. "Kazi Nazrul Islam: The Voice of Poetry and the Struggle for Human Wholeness". *Nazrul Institute, Kabi Bhaban, Dhaka, Bangladesh*

Zuckerman, Phil. *Society without God: What the Least Religious Nations Can Tell Us about Contentment.* New York University Press, 2008. Cited in http://en.wikipedia.org/wiki/secular_morality.

————. *Living the Secular Life: New Answers to Old Questions.* New York: Penguin Press, 2014.

ENDNOTES

Introduction

1 Source: http://www.ajol.info/index.php/lwati/article/view.
2 Source: www.livingwithoutreligion.com.
3 Michael Shermer, 2002, *Why People Believe Weird Things*, in *Why I Became an Atheist*, John W. Loftus (New York: Prometheus Books, 2012).
4 Marcus, 2000, in *Why I Became an Atheist*, John W. Loftus (New York: Prometheus Books, 2012).
5 Carol Tavris and Elliot Aronson, 2007, in *Why I Became an Atheist*, John W. Loftus (New York: Prometheus Books, 2012).
6 "Humanism and Religion or How to Thread a Needle," by Marilyn Westfall, www.google.ca/Humanism+and+Religion+How+to+Thread+a+Needle+by+Marilyn+Westfall.
7 Source: http://humanityplus.org/philosophy/.
8 Source: www.from-humanism-to-transhumanism.
9 Source: http://blogs.discovermagazine.com/sciencenotfiction/2011/07/16/when-will-we-be-transhuman.
10 Seung, Sebastian, *Connectome: How the Brain's Wiring Makes Us Who We Are* (Mariner Books, 2013), in "Transhumanism and the Meaning of Life," Anders Sandburg, http://www.aleph.se/papers/Meaning%20of%20life.pdf.

Chapter One: Religion

11 Source: "Evolutionary Origin of Religions," *Wikipedia*.
12 Source: "The Last Two Million Years," *The Reader's Digest*, 1974.
13 Cited from "The Biological Roots of Religion: Is Faith in Our Genes?" http://www.secularhumanism.org/library/fi/hunt_10_3.html.
14 Morton Hunt, http://www.secularhumanism.org/library.

15 Ibid.
16 Cited from "The Neurological Origins of Religious Belief,"
    by John Cookson, http://bigthink.com/going-mental/
    the-neurological-origins-0f-religious-belief.
17 Ibid.
18 Ibid.
19 Ibid.
20 Source: http://thehumanist.com/magazine/
    november-december-2013/arts_entertainment/
    the-religion-virus-why-we-believe-in-god-2.
21 Ibid.
22 Cited from "A Psychological Perspective on the Source and Function
    of Religion," http://www.hts.org.za/index.php/HTS/rt/.
23 Source: http://www.infidels.org/library/historical/robert_ingersoll/
    what_is_religion.html
24 Ibid.
25 "Albert Einstein on Religion and Science." Cited from http://www.
    sacred-texts.com/aor/einstein/einsci.htm.
26 "The Universe in a Nutshell" by Stephen Hawking. Cited from
    http://www.hawking.org.uk/lecture/dice.
27 Source: http://en.wikipedia.org/wiki/Ethics/Spinoza.
28 Ibid.
29 Source: http://www.quora.com/Philosophy-of-Religion/
    Does-the-god-of-Spinoza.
30 Source: https://en.wikipedia.org/wiki/Spinozism.
31 Source: http://rationalwiki.org/wiki/Baruch_Spinoza.
32 Source: https://whistlinginthewind.org/2012/05/22/
    religion-as-a-cause-of-wa.
33 Source: http://www.thetoptens.com/
    atrocities-committed-name-religion/.
34 Source: http://www.religioustolerance.org/curr_war.htm.
35 Source: http://www.ppu.org.uk/learn/infodocs/st_religions.html.
36 Ibid.
37 Source: http://www.religioustolerance.org/curr_war.htm.
38 Source: http://www.theguardian.com/world/2015/mar/27/
    israel-kills-more-palestinians-2014-than-any-other-year-since-1967.
39 Source: http://garryleech.com/2014/08/08/
    israel-palestine-by-the-numbers/.
40 Source: https://en.wikipedia.org/wiki/Honor_killing.
41 Source: http://www.justice.gc.ca/eng/rp-pr/cj-jp/fv-vf/hk-ch/
    p3.html.

42 Source: http://www.ibtimes.co.uk/ india-pakistan-account-2000-honour-killings-every-year.
43 Source: http://www.jihadwatch.org/2014/07/ in-canada-defending-girls-from-islamic-honor-killings.
44 Source: http://www.meforum.org/3287/ hindu-muslim-honor-killings.
45 Source: http://www.violenceisnotourculture.org/content/ education-key-prevent-honor-killings.
46 Source: http://timesofindia.indiatimes.com/india/ More-than-1000-honour-killings-in-India-every-year-Experts.
47 Source: http://www.foxnews.com/us/2015/11/10/ honor-killing-in-us-justice-department.
48 Source: https://en.wikipedia.org/wiki/Honor_killing.
49 Source: http://freethoughtblogs.com/taslima/2012/04/17/ crimes-against-humanity/.
50 Source: http://www.hrw.org/.
51 Source: http://www.unfpa.org/child-marriage.
52 Source: http://www.girlsnotbrides.org/about-child-marriage/.
53 Source: http://terredasie.com/english/english-articles/ history-of-child-marriage-in-india/.
54 Source: https://www.icrw.org/files/publications/Child_marriage/ Asia.2013.
55 Source: http://www.girlsnotbrides.org/child-marriage/.
56 Source: http://www.un.org/youthenvoy/2013/09/child-marriages-39000-every-day-more-than-140-million-girls-will-marry-between-2011-and-2020/.
57 Source: http://www.truthbeknown.com/introduction.htm.
58 Source: http://www.forwardprogressives.com/ humanity-will-always-live-chaos-long-organized-religion-persists/.
59 Source: http://www.alternet.org/ belief/6-ways-religion-does-more-bad-good.
60 Ibid.
61 Ibid.
62 http://www.amazon.com/ Living-Secular-Life-Answers-Questions-ebook/dp.
63 http://redcresearch.ie/wp-content/uploads/2012/08/RED-C-Press-release-Relogion-and-Atheism-25-7-12.pdf.
64 http://www2.psych.ubc.ca/-ara/IndexBook.html.
65 "Memes of Ethics—A Co-Evolutionary Approach. The Case of Religion's Memes." K. E. Simitopoulou and N. I. Xirotiris. Cited

from krepublishers.com/.../JHE- SI-12-04-023-027-Simitopoulou-K-Text.pd.

66 Source: http://www.123helpme.com/search.
    asp?text=organized+religion.

67 "Theory of Religion as Myth: On Loyal Rue (2005),
    Religion Is Not about God" by Hubert Seiwert, https://
    www.gko.uni-leipzig.de/fileadmin/userupload/
    religionswissenschaft/Pdf/Publikationen_Seiwert/
    Seiwert_-_Theory_of_religion_as_myth.pdf.

68 "Current Trends in the Theories of Religious Studies: A Clue to
    Proliferation of Religions Worldwide" by Nathaniel Aminorishe
    Ukuekpeyetan—Agbikimi, http://www.eajournals.org/wp-content/
    uploads/Current-Trends-in-Theories-of-Religious-Studies.pdf.

69 Ibid.

70 Ibid.

71 Ibid.

72 Ibid.

73 Ibid.

74 Ibid.

75 Source: http://hirr.hartsem.edu/ency/Anthropology.htm.

76 Source: https://en.wikipedia.org/wiki/Anthropology_of_religion.

77 Source: http://www.iawwai.com/UnificationOfReligion.htm.

78 Source: http://reasonandmeaning.com/2015/04/30/
    does-morality-depend-on-religion-answered-in-two-pages/.

79 Source: https://prezi.com/-vounxznphgc/
    does-morality-depend-on-the-existence-of-relgion/.

80 Source: http://atheistexperience.blogspot.com/20...

81 Source: https://answers.yahoo.com/question/index?qid.

Chapter Two: Beyond Religion

82 Source: http://www.dailymail.co.uk/news/article-2250096/You-
    wouldnt-believe-atheism-worlds-biggest-faith-Christianity-Islam.
    html#ixzz3YdaZ3Mpd.

83 Source: http://www.christianhumanist.net/confession-of-21st-
    century-christian.html.

84 Source: http://testimonials.exchristian.net/2003/10/why-i-reject-
    christianity.php.

85 Source: http://freethoughtblogs.com/greta/2012/01/05/christopher-
    hitchens-brother-atheists-reject-religion-so-they-can-be-decadent/.

86 Source: http://freethoughtblogs.com/greta/2011/06/07/
    do-atheists-have-better-sex/.

87 Source: http://en.wikipedia.org/wiki/Atheism.
88 Source: http://www.humanreligions.info/forces.html#Intro.
89 Source: https://www.vocabulary.com/dictionary/pantheism.
90 Source: www.iawwai.com/Bibliography.htm.
91 Source: http://pro-prosperity.com/A-Unifying-Philosophy-of-Governance.html.
92 Ibid.
93 Source: https://consortiumnews.com/2015/05/03/letting-scientific-knowledge-into-religion/.
94 Rev. A. Powell Davies, DD, http://dmuuc.org/aboutworship/dr-a-powell-davies-bio-sermons/can-science-religion-together/.
95 Source: https://moralmusing.wordpress.com/2013/01/03/elements-of-moral-philosophy-does-morality-depend-on-religion/.
96 Source: http://www.anti-naturals.org/theory/religion.html.
97 "God, the Failed Hypothesis: How Science Shows That God Does Not Exist"by Prof. Victor J. Stenger (2007) http://www.humanreligions.info/secular_morals.html.
98 "Religion Explained" by Pascal Boyer, 2001, http://www.humanreligions.info/secular_morals.html.
99 Source: http://en.wikipedia.org/wiki/Secular_morality.
100 Source: http://www.stevepavlina.com/blog/2008/05/10-reasons-you-should-never-have-a-religion/
101 Ibid.
102 Source: https://www.psychologytoday.com/blog/unique-everybody-else/201401/more-knowledge-less-belief-in-religion.
103 Source: http://reason.com/archives/2016/03/25/science-is-a-good-substitute-for-god.
104 Source: http://www.patheos.com/blogs/progressivesecularhumanist/2016/03/report-worlds-happiest-countries-are-also-least-religious/.

Chapter Three: Humanism
105 Source: http://en.wikipedia.org/wiki/Humanism.
106 Source: https://www.bu.edu/wcp/Papers/Reli/ReliMeht.htm).
107 Source: https://humanism.org.uk/humanism/the-humanist-tradition/renaissance/.
108 Source: http://humanist.org.sg/humanism/introduction-to-humanism/history-of-humanism/.
109 Source: https://humanism.org.uk/humanism/the-humanist-tradition/renaissance/.

110 Source: http://en.wikipedia.org/wiki/Humanism.

111 Source: www.thehumanistparty.com.

112 Source: http://www.lfpress.com/2015/01/23/emerson-humanism-essential-to-good-religion.

113 Source: http://iheu.org/humanism/what-is-humanism/.

114 Source: http://david-pollock.org.uk/humanism/humanism-beliefs-and-values/.

115 Ibid.

116 Source: http://secularhumanism.blogspot.ca/2006/12/meaning-of-life.html.

117 Source: http://en.wikipedia.org/wiki/Religious_humanism.

118 "Reason and Reverence: A New Religious Humanism" by Rev. Dr. William R. Murry, cited from http://www.nysec.org/uploads/nysectalk080921_2.pdf.

119 Source: https://www.secularhumanism.org/index.php.

120 Source: http://www.christiananswers.net/q-sum/sum-r002.html.

121 Source: http://instituteforscienceandhumanvalues.com/articles/neo-humanist-statement.htm.

122 Source: http://end-blasphemy-laws.org/2015/01/lets-abolish-all-blasphemy-laws-worldwide/.

123 Source: http://instituteforscienceandhumanvalues.com/articles/neo-humanist-statement.htm.

124 Source: http://www.transcript-verlag.de/media.

125 Source: http://www.pluralism.org/religion/humanism/belief.

126 Source: https://www.quora.com/Do-you-think-humanism-is-important-to-to-the-future-of-the-human-species.

127 Source: http://thehumanist.com/magazine/november-december-2015/features/humanism-doubt-and-optimism.

128 Ibid.

129 Ibid.

130 Introductory talk to the newly formed East London Humanist Group, 24/9/12, by Zelda Bailey.

131 Source: http://www.huffingtonpost.com/phil-zuckerman/imagine-no-religion.

132 Source: http://www.santabarbarahumanists.org/humanism-and-its-aspirations.

133 Source: http://americanhumanist.org/humanism/definitions_of_humanism.

134 Ibid.

135 Source: http://www.thinkhumanism.com/the-golden-rule.html.

136 Source: http://humanistinstitute.org/about-us/defining-humanism/.

137 Source: http://instituteforscienceandhumanvalues.com/articles/neo-humanist-statement.htm.

Chapter Four: Transhumanism
138 Source: http://jp.senescence.info/thoughts/transhumanism.html.
139 Source: http://rationalwiki.org/wiki/Transhumanism.
140 Ibid.
141 Ibid.
142 Kristi Scott, 2011, http://ieet.org/index.php/IEET/more/scott20110714.
143 Source: http://www.livescience.com/45872-transhuman-technology.html.
144 Source: http://io9.gizmodo.com/5967896/us-spy-agency-predicts-a-very-transhuman-future-by-2030.
145 Source: http://en.wikipedia.org/wiki/Transhumanism.
146 "Transhumanism and the Meaning of Life," Anders Sandburg, cited from http://www.aleph.se/papers/Meaning%20of%20life.pdf.
147 Source: https://lifeboat.com/ex/transhumanist.technologies#1.
148 Ibid.
149 Ibid.
150 Ibid.
151 Ibid.
152 Ibid.
153 Ibid.
154 Ibid.
155 Ibid.
156 Ibid.
157 Source: http://americanhumanist.org/HNN/details/2013-09-movie-review-elysium-transhumanism-and-the-relations.
158 Source: http://whatistranshumanism.org/.
159 Source: http://blogs.discovermagazine.com/sciencenotfiction/2011/07/16/when-will-we-be-transhuman-seven-conditions-for-attaining-transhumanism.
160 Source: http://www.huffingtonpost.com/zoltan-istvan/are-you-ready-for-the-fut_b_10199682.html.
161 Source: http://www.dailydot.com/via/zoltan-istvan-rfid-chip-implant/.
162 Source: http://www.thenanoage.com/transhumanism-posthumanism.htm.

Chapter Five: Summary and Conclusion

163 Source: https://en.wikipedia.org/wiki/ Relationship_between_religion_and_science.

164 Source: https://moralmusing.wordpress.com/2013/01/03/ elements-of-moral-philosophy-does-morality-depend-on-religion/.

165 Source: http://www.anti-naturals.org/theory/religion.html.

166 Source: http://en.wikipedia.org/wiki/Humanism.

167 Source: https://www.bu.edu/wcp/Papers/Reli/ReliMeht.htm.

168 Source: https://humanism.org.uk/humanism/ the-humanist-tradition/renaissance/.

169 Ibid.

170 Source: http://humanist.org.sg/humanism/ introduction-to-humanism/history-of-humanism/.

171 Source: https://humanism.org.uk/humanism/ the-humanist-tradition/renaissance/.

172 Source: http://en.wikipedia.org/wiki/Humanism.

173 Source: www.thehumanistparty.com.

174 Source: http://iheu.org/humanism/what-is-humanism/.

175 Source: http://david-pollock.org.uk/humanism/ humanism-beliefs-and-values/.

176 Ibid.

177 Source: http://secularhumanism.blogspot.ca/2006/12/meaning-of-life.html.

178 Source: http://www.pluralism.org/religion/humanism/belief.

179 Flemin Rod, https://www.quora.com/Do-you-think-humanism-is-important-to-to-the-future-of-the-human-species.

180 Hancook, Jennifer, https://www.quora.com/Do-you-th. nk-humanism-is-important-to-to-the-future-of-the-human-species.

181 Source: http://thehumanist.com/magazine/ november-december-2015/features/humanism-doubt-and-optimism.

182 Ibid.

183 Source: www.cordeiro.org.

184 Source: http://rationalwiki.org/wiki/Transhumanism.

185 Source: http://www.livescience.com/45872-transhuman-technology. html.

186 Source: http://en.wikipedia.org/wiki/Transhumanism.

187 Source: https://aeon.co/essays/ what-does-it-take-to-make-a-stand-for-free-speech.

188 Source: https://dawnofgiants.wordpress.com/tag/transhumanist/.

189 Source: https://aeon.co/essays/ at-a-time-of-zealotry-spinoza-matters-more-than-ever.

INDEX

www.ingramcontent.com/pod-product-compliance
Lightning Source LLC
Chambersburg PA
CBHW020727180526
45163CB00001B/144